GET YOUR TEETH INTO IT!

TEACHING FRAMEWORK
Year 5

Series Editor: Jane Turner

Series Consultant: Derek Bell

Authors:

Chris Banbury, Nicola Beverley, Hellen Ward

William Collins' dream of knowledge for all began with the publication of his first book in 1819. A self-educated mill worker, he not only enriched millions of lives, but also founded a flourishing publishing house. Today, staying true to this spirit, Collins books are packed with inspiration, innovation and practical expertise. They place you at the centre of a world of possibility and give you exactly what you need to explore it.

Collins. Freedom to teach.

Published by Collins
An imprint of HarperCollins*Publishers*
77–85 Fulham Palace Road
Hammersmith
London
W6 8JB

Browse the complete Collins catalogue at
www.collins.co.uk

© HarperCollins*Publishers* Limited 2014

10 9 8 7 6 5 4 3 2 1

ISBN 978-0-00-755145-3

The authors assert their moral rights to be identified as the authors of this work

British Library Cataloguing in Publication Data

A Catalogue record for this publication is available from the British Library

Authors: Chris Banbury, Nicola Beverley, Hellen Ward
Contributing Author: Christine Moorcroft
Publishing Manager: Lizzie Catford
Managing Editor: Helen Terrington
Development Editor: Melanie Hoffman
Project Editors: Mike Appleton, Alison Walters
Production: Rebecca Evans
Editors: Soo Hamilton, Sophia Ktori, Joan Miller, Christine Moorcroft, Sarah Vittachi
Cover design and artwork: Amparo Barrera
Internal design concept: Amparo Barrera
Designer: Linda Miles, Lodestone Publishing Limited
Illustrations: Aptara Inc., Ian Moores, Mike Lacey

Printed and bound by Martins the Printers Ltd, Berwick-upon-Tweed

FSC is a non-profit international organisation established to promote the responsible management of the world's forests. Products carrying the FSC label are independently certified to assure consumers that they come from forests that are managed to meet the social, economic and ecological needs of present and future generations, and other controlled sources.

MIX
Paper from
responsible sources
FSC™ C007454

Find out more about HarperCollins and the environment at **www.harpercollins.co.uk/green**

Health and safety: The publishers, series editor and authors have made every reasonable effort to ensure that the activities in this book are safe when conducted as instructed. However, the publishers, series editor and authors assume no responsibility for any damage or injury caused or sustained while performing the activities in this book to the full extent permitted by the law. Parents, guardians and/or teachers should supervise pupils who undertake the activities in this book.

CONTENTS

INTRODUCTION

WELCOME TO SNAP SCIENCE!

The publication of the new National Curriculum for Science in Primary Schools in England is a wonderful opportunity for subject leaders and teachers to review all aspects of their science provision, including planning, pedagogy, resourcing and assessment. There are several new topics to get to know, a wider range of science enquiry must be planned for, children will need to spend more time learning science outdoors and assessment will no longer be about 'levelling'. All these changes are undoubtedly positive, yet they have big implications for busy, dedicated primary teachers. How can you be sure that the transition to the new Programmes of Study for Science for Key Stages One and Two in your school is smooth and stress free and, most importantly, results in enjoyable, challenging and successful science learning for all children?

Snap Science has been created by a team of leading experts to give subject leaders and teachers the confidence to develop a new scheme of work for science for their school which clearly meets the aims of the new National Curriculum for Science. It covers all the required knowledge, conceptual understanding and the full range of scientific enquiry types identified in the Programmes of Study. Snap Science is a comprehensive and rich resource that will support best practice in science teaching and teacher assessment, whilst encouraging teacher professionalism and autonomy. It has been written by teachers for teachers with full awareness that every class is different but that teachers share similar concerns:

• Am I meeting the requirements of the Programmes of Study?
• Is there clear progression in the science learning in my school?
• Is the level of challenge right for each child in my science lessons?
• Is my subject knowledge secure for all topics?
• Is formative assessment built into every lesson?
• Am I using the best teaching strategies for this topic?
• Are children engaged in productive practical work and meaningful enquiry?
• Do children enjoy their science learning?
• Are the children in my class making good progress in science?

When using Snap Science you can be confident that these concerns are addressed.

Snap Science comprises:

• An online resource kit with a flexible planning tool, editable lesson plans, integrated assets for every lesson (including videos, animations and slideshows), as well as support for assessing and tracking children's progress.
• A printed Teaching Framework per year group with sequenced lesson plans for each topic in the new Programme of Study.

Together these form a tool kit which has been designed to enable and liberate teachers to do what they do best – teach science well! I hope you enjoy using Snap Science.

Jane Turner

THE BEST TEACHING OF PRIMARY SCIENCE

Snap Science has been created to reflect current ideas about best practice in primary science teaching and learning. Although a relatively new subject in primary schools in England, becoming statutory only in 1988, science is a well understood area of the Primary Curriculum. There is a wealth of research evidence and good practice data that has informed the design and content of Snap Science. At the heart of the resource are the following principles:

1 A science scheme of work must embody a clear progression.

2 It is through working scientifically that children develop an understanding of the nature and processes of science and the key scientific knowledge and concepts.

3 Children are curious to find answers to questions about the world around them.

4 Children need to be actively involved in their own learning, to be engaged and reflective.

5 Every child should have the opportunity to achieve in every lesson.

6 Assessment is an integral part of teaching that enables children to understand the purpose of their activities and to improve the quality of their work.

These principles have shaped the overall design of Snap Science, as well as the content and structure of each lesson.

1 Based on the year by year model of the Programmes of Study, the creators of Snap Science have developed a **clear progression framework of the 'big ideas in science'** which has been used to structure the content both within each topic, from year to year and within each year group and module, and to identify any conceptual gaps. This will ensure that children are continually building on their prior learning as they systematically develop their understanding of key ideas and their scientific skills.

2 The creators of Snap Science recognise that **working scientifically**, asking questions and testing ideas against evidence, is the most effective way for children to learn about science. Therefore each lesson **has a clear science enquiry focus**.

3 Every lesson in Snap Science is carefully planned around **a question for children to answer**, either inside the classroom or outside. By ensuring that these questions spark children's curiosity and that they want to find out the answer, lessons are purposeful and result in children gaining a new understanding of the world around them.

4 In each lesson in Snap Science the **learning intention** is designed so that children have a powerful understanding of the skills and understanding they are developing in the lesson. **Success criteria** define the features of the learning intention in the context of the activity so that children can identify what they are aiming for and how well they are doing.

5 Snap Science has been designed to ensure that all children in a class can access and master the lesson's learning intention with each lesson offering **three levels of differentiated task**. These are planned to challenge and extend the learning of all children whilst ensuring that they all achieve the learning intention.

6 Every lesson in Snap Science includes **Assessment for Learning strategies** which enable teachers to find out what children have learned and to use that information formatively.

7 In response to the wealth of evidence that exists about the benefits of children experiencing the natural world first hand, **children learning science outdoors** is a key feature of Snap Science. For each year group there is a module called **Our Changing World** which is designed to be taught in every term, offering children regular opportunities to explore all aspects of their outdoor environment and build up a rich understanding of how it changes over the year.

The most recent Ofsted report[1] into science in primary schools in England outlined three factors which exemplify the best teaching:

- It is driven by determined subject leadership that puts **scientific enquiry at the heart of science teaching** and coupled with **substantial expertise** in how pupils learn science
- It sets out to sustain **pupils' natural curiosity**, so that they are eager to learn the **subject content** as well as develop **the necessary investigative skills**
- It is informed by **accurate and timely assessment** of how well pupils are developing their understanding of science concepts and their skills in analysis and interpretation so that **teaching can respond to and extend pupils' learning**.

The creators of Snap Science have considerable and acknowledged expertise in how children in primary schools learn science well. This experience and knowledge has enabled them to design a resource which embodies the best of current pedagogy and practice, and meets Ofsted's definition of 'the best teaching'.

SCIENCE ENQUIRY

A scheme of work that has **science enquiry at its heart** requires children to learn to use a variety of approaches to answer relevant scientific questions[2]. Each module in Snap Science is made up of a carefully planned series of lessons which will engage children in **the different types of science enquiry identified in the National Curriculum**, where they will use and develop **the necessary investigative skills** and attributes identified for each Key Stage phase. All lessons are stimulated by a question for children to answer, a scientific phenomenon to investigate or a problem to solve. Science Enquiry is the methodology children will use to develop their conceptual knowledge, working in an authentically scientific and purposeful way to collect evidence to find answers to their questions.

SUBJECT KNOWLEDGE CONFIDENCE

Snap Science has been carefully planned to ensure that children use their developing science enquiry skills to build their knowledge of the scientific ideas in a systematic and conceptually appropriate way. Each lesson is designed to explore, value and build on children's prior knowledge so that misconceptions can be addressed and secure understanding developed. The creators of Snap Science are aware that every primary teacher cannot be an expert in all subjects, and have designed the resource to give teachers **subject knowledge confidence**. The sequence of science ideas throughout the resource as a whole, and in each module and lesson, is clear and accurate. The introduction to each module provides teachers with a clear explanation of the science they need to understand and in each lesson key information is highlighted at relevant points.

EXPLORE ACTIVITIES

Stimulating and maintaining **children's natural curiosity** is fundamental to good science teaching and learning. Every lesson in Snap Science starts with an **Explore activity** to excite children's curiosity about a scientific phenomenon and provide a focus for their questions and investigations. The Explore activity is also designed as a **rich formative assessment opportunity** for children to reflect on what they already know, and identify what they need to learn next.

---

[1] Ofsted 2013 Maintaining Curiosity

[2] Goldsworthy, A., Watson, R., Wood-Robinson,V. (2000) *Investigations: Developing Understanding* Hatfield: Association for Science Education

Turner, J., Keogh, B., Naylor, S., Lawrence, L., (2011) *It's Not Fair - Or Is It?* Sandbach: Millgate House Publishers and Association for Science Education

ENQUIRE CHALLENGES

Genuine curiosity leads to **authentic, purposeful science investigation** so each Explore activity is followed by a **differentiated Enquire challenge** where children will collect and analyse data to answer their questions and so develop their scientific understanding and knowledge. Mostly children's science investigations will involve them in **first hand collection, recording and analysis of data**, although sometimes they will use secondary sources of evidence or information to answer questions. In the Enquire challenges children will engage in a **wide range of practical activity** both indoors and outside, using a variety of observation and measuring equipment including data loggers and digital microscopes, everyday items and materials, natural and living things and electrical components. Comprehensive equipment lists are supplied for every module and lesson to help teachers with planning and ensure that **children have as much independence as possible to decide what data to collect to answer their question and how**. Snap Science supports children to work both in groups and alone as appropriate.

SUMMARISE, SHARE, REFLECT

Good science teaching recognises that children need opportunities to **summarise what they have found out, share their findings and reflect on what they have learned**. Each Snap Science lesson has a final **Reflect and Review** activity when children will communicate what they have learned in an appropriate and meaningful way. Being able to summarise understanding is key to developing conceptual knowledge as well as being the vital, final satisfying step in the science enquiry process. **Writing, drawing, speaking, using ICT and mathematical** formats are all important skills in communicating and presenting science and are all developed in Snap Science.

The Reflect and Review part of each lesson also provides an excellent opportunity for children to **self and peer assess their achievements** in the lesson or module, using the success criteria to guide them. How well have I completed the challenge? What do I know now that I didn't know at the beginning of the lesson? What have I learnt to do? What can I do better now than I could at the beginning of the lesson? What do I want to find out next? What do I need to do next to improve the skills I used today?

As children reflect on their own learning, teachers can also assess the progress that they are making. Each lesson in Snap Science includes **guidance for teacher assessment**, indicating where teachers will find evidence of achievement of the learning intention and what that achievement may look like, in the things that children say, do, write or draw. Assessment will therefore be on-going and accurate – focusing firmly on progress in conceptual knowledge plus data collection and analysis skills. Assessment will also be formative, supporting teachers and children to identify the next steps in learning, and to keep moving forward.

THE NEW PROGRAMMES OF STUDY FOR SCIENCE

In September 2013 the Department for Education published a new Primary National Curriculum including a new Programme of Study for Science. Implementation begins September 2014 with the first sample tests reporting on national standards against the new POS in 2016.

What is different about the new National Curriculum for Science? What challenges and opportunities do these changes present to primary teachers? How will Snap Science help teachers to meet them?

As with any curriculum it is important to know and understand its aims. It is the aims that explain what the education described should do. In the case of the new National Curriculum for Science the aims are reflective of widely and long held views about the values and purposes of primary science education.

The National Curriculum for Science aims to ensure that all children:

1 Develop **scientific knowledge and conceptual understanding** – through the specific disciplines of biology, physics and chemistry,

2 Develop understanding of the **nature, processes and methods of science** through different types of science enquiry that help children to answer scientific questions about the world around them, and

3 Are equipped with the scientific knowledge required to understand the **uses and implications** of science, today and for the future.

1 The Programmes of Study contain a **sequence of knowledge and concepts** on a year by year basis. Although it is not compulsory for schools to follow this sequence, the blocks of knowledge and concepts have been arranged to ensure progress in the big ideas of science to ensure that children develop secure understanding. Most of the content will be familiar to teachers, but some topics have been broadened and extended, and some are introduced at different times.

- The plant and animal biology content is significantly increased for every year group, with much more focus on children getting to know and to classify living things in their local and wider outdoor environment.
- A wider range of plant and animal life cycles is included.
- Evolution and inheritance is now included at Y6.
- Simple digestion in humans is included at Y3.
- Seasonal changes is now included at Y1 including day length.

- Chemistry at Y3 now includes fossils.
- States of matter is now in Y4, mixing and changing materials is all in Y5.
- There is no chemistry content specified for Y6.

- These is no requirement to cover electricity, light and sound, and forces and movement in KS1.
- Mechanisms are included at Y6.

To enable teachers to confidently plan to teach the new National Curriculum, Snap Science is organised into a series of modules per year group, based on the topics in the new Programmes of Study, and the year groups in which they occur. All the required science content is fully covered to make sure that learning is deep. Each module is made up of a sequence of lessons which has been carefully planned by subject experts to ensure that new ideas are only introduced once understanding of lower-order content is secure. Although in most lessons children will engage in first hand practical activity, each lesson is also richly supported by additional assets including film clips, images and interactive resources to exemplify and illustrate concepts and ideas.

Because the creators of Snap Science know that primary schools are organised in many different ways the interactive planning tool will facilitate flexible whole school planning, whilst still ensuring coherent progression of knowledge and understanding.

2 In the new National Curriculum **the nature, processes and methods of science** are organised into three sections for KS1, lower and upper KS2 under the title **working scientifically**. These sections of the Programmes of Study identify the progress children should make in all areas of science enquiry from asking questions, planning and carrying out investigations, presenting data, making and communicating conclusions and evaluating results.

WORKING SCIENTIFICALLY

However a difference in the new Programmes of Study is that working scientifically, although described separately, is not to be taught as a separate strand. The creators of Snap Science fully support this embedded approach as it reflects accurately the scientific approach and ensures that science lessons are purposeful and lead to the learning of science concepts. Each lesson in Snap Science is planned to meet a biology, chemistry or physics learning intention by Working Scientifically. Teachers will therefore be able to track progression in all aspects of the Programmes of Study.

The new Programmes of Study require that children should learn to use a variety of approaches to answer scientific questions, as well as fair testing which has become an over and often incorrectly used method in primary classrooms[3]. The creators of Snap Science know that different questions lead to different types of enquiry. Teachers can be confident that the starting points in Snap Science will require children to use the recommended different types of science enquiry to answer questions and so develop their understanding about which is the best method to use to answer a question. In each lesson in Snap Science the enquiry strategy that children will use is clearly identified. Enquiry strategies include those recommended in the new National Curriculum:

- **Observing over time** – when children observe or measure how one variable changes over time
- **Identifying and classifying** – when children identify and name materials and living things and make observations or carry out tests to organise them into groups
- **Looking for patterns** – when children make observations or carry out surveys of variables that cannot be easily controlled and look for relationships between two sets of data
- **Comparative and fair testing** – when children observe or measure the effect of changing one variable when controlling others as far as possible
- **Answering questions using secondary sources of evidence** – when children answer questions using data or information that they have not collected first hand

As well children will:

- **Use models** – to develop or evaluate a model or analogy that represents a scientific idea, phenomenon or process

---

[3] Turner, J., Keogh, B., Naylor, S., Lawrence, L., (2011) *It's Not Fair – Or Is It?* Sandbach: Millgate House Publishers and Association for Science Education

3 Teachers sometimes find daunting the idea of helping children to understand the **uses and implications of science**, citing their own lack of confidence about complex scientific developments and issues. The creators of Snap Science recognise that this aim is challenging and have developed a straightforward two-fold strategy to support teachers to develop children's scientific literacy.

- Firstly the **context of the lesson must make sense and matter to children**. They need to see the relevance of a scientific question or concept to their own lives. In every lesson children should be able to explain the importance of the question they are answering, and how the science connects to their own lives. In Snap Science the approach to this is pragmatic and manageable, using question and stimulus Explore activities to focus children's powerful curiosity about the world around them. At the end of every lesson in Snap Science children should be able to explain or demonstrate how they have answered the question and what they have learnt. Sometime children will do this via a **technology** activity – applying their scientific knowledge and understanding to make an artefact or system that solves a problem. Sometimes they will use argument, debate or persuasive writing to show how a science explanation has helped them to understand or make informed personal decisions about something that involves science, such as health, diet, use of energy resources, or human impact on the natural environment.

- Secondly children must learn that all **understanding in science depends on the evidence that has been used to answer a question** and that working scientifically involves evaluating the quality of that evidence and the conclusions that have been drawn from it. In every lesson children should be able to explain what evidence they have used to answer a question and to evaluate honestly the reliability and validity of that evidence. In Snap Science children are supported to use discussion and argument to evaluate their own data, methods and conclusions, as well as those of others, in the classroom and beyond. They will also be helped to recognise how improvements in evidence collection and analysis techniques have led to ideas about science changing over time.

ASSESSING THE NEW NATIONAL CURRICULUM FOR PRIMARY SCIENCE

Alongside new Programmes of Study for primary science, a new assessment model will be introduced, where attainment will no longer be tracked and reported against level descriptors, but instead children's 'mastery' of the matter, skills and processes of the Programme of Study will be assessed. The expectation is that most children will achieve 'mastery' of the full programme of study. Sample tests in Science to provide national monitoring of standards will take place every two years, but involving a small number of pupils, with no individual pupil or school Science attainment data being reported. Schools are now free to decide how to track the progress that children make through KS1 and 2 against the Programmes of Study.

What are the implications of removing levels in primary science? What challenges and opportunities does this change present to primary teachers? How will Snap Science help teachers to meet them?

Levels, particularly when used to design SAT questions, had the effect of constraining science learning in two ways: firstly the curriculum became narrowed to what could be tested in a replicable, pen and paper method and therefore understanding of what achievement looked like came to be understood in terms of these narrow tasks; and secondly they became organising models for planning, with teachers differentiating lessons according to artificial notions of children 'levelness'. [4]

---

[4] Harlen (2012), Developing Policy, Principles and Practice in Primary Science Assessment: Report from a Working Group; London: Nuffield Foundation

The removal of level descriptors means that the **relationship between the science that children are taught and the science that is assessed will be much stronger**. Confident teacher assessment is vital and the creators of Snap Science have ensured that **effective formative assessment strategies** are used in every lesson.

- Each lesson has a clear science **Learning Intention** which all children are expected to achieve or exceed, with **Success Criteria** to exemplify what success will look.

- Differentiation is by access, with each lesson beginning with an **Explore** activity to enable children and their teacher to assess prior understanding and identify which level of challenge to take. Teachers can annotate planning to reflect this.

- The **Enquire** part of the lesson includes a choice of three challenges which will ensure that all children can work appropriately towards achieving the learning intention. Differentiation in the challenges is based on a model of progression in science learning which supports children to become more independent and autonomous, systematic, precise and evaluative, and to increasingly use their scientific knowledge in their explanations. This means that grouping in Snap Science lessons is flexible, dependent on the level of skill, knowledge and understanding that each child demonstrates in the Explore activity, and the level of support and challenge that is appropriate for them in each lesson. Children should be encouraged, with teacher support, to choose for themselves the right challenge to complete to achieve the learning intention.

- The final stage of each lesson is a **Reflect and Review** activity where children summarise what they have learnt and use the success criteria to assess their success and identify next steps. Assessment evidence from each lesson should be used formatively to determine appropriate next steps for individuals and groups of children.

Without levels to track progress against teachers will need to find other ways of monitoring children's progress and reporting this to parents, secondary schools and external bodies such as Ofsted. The creators of Snap Science recognise that this represents a challenge to schools and have referred to recognised sources of good practice[5] to design a manageable process for tracking progress in science.

- **Formative assessment evidence from each lesson,** including children's work, the feedback that is given and responded to and any additional observation notes that the teacher makes, is used to track progression and to enable teachers to make confident summative judgments of attainment when required. Comparison of evidence outcomes from different children and between classes and school is used to **moderate teacher assessment judgements**.

- Supporting digital assessment tasks on Collins Connect such as quizzes or short activities can be used to **check children's understanding** of specific concepts or facility with particular working scientifically skills.

- Either at an identified point during a module or at the end of a module a teacher reviews any observation notes on a child, their written work, their self-assessment judgements, and their answers to any additional activities they have completed to ascertain if they **have not yet achieved, have achieved, or have achieved and exceeded the expected outcomes** for that part of the Programme of Study. Teachers **record the judgements in an on-going digital progress tracker on Collins Connect** which contains a summary of the knowledge outcomes for that module including an on-going Working Scientifically tracker.

- There are no 'end of module tests' nor artificial interim (sub) levels of achievement. It is assumed that high quality formative assessment which influences planning for individual children will lead

---

[5] Harlen (2012), Developing Policy, Principles and Practice in Primary Science Assessment: Report from a Working Group; London: Nuffield Foundation

to an excellent match of task and challenge and that teachers and children will recognise when the lesson intention has been achieved.

- At the end of each Key Stage teachers use the digital progress trackers on Collins Connect to judge **whether or not each child has achieved the designated learning outcomes for the Key Stage** in the main components of the National Curriculum.

- At the end of the Key Stage **individual pupil records are aggregated from each class,** for the school as a whole and for particular groups.

SNAP SCIENCE COMPONENTS

THE TEACHING FRAMEWORK

- The printed Teaching Framework is organised into a series of modules, based on the topics in the new Programmes of Study. The new Programme of Study for Science is divided into 4 topics per year in KS1 and 5 per year in KS2. Some topics have considerably more content than others. Traditional half termly topic planning for science will clearly no longer be appropriate. The modules in Snap Science have been organised to reflect the content of the POS for that year and are not all of identical length. Teachers should plan to teach all the complete modules over a year, regularly fitting in lessons from the Our Changing World module.

- Each module begins with an introduction that provides background information on the topic at an appropriate level for non-specialist teachers as well as advice on the misconceptions or alternative conceptions that research indicates that children frequently develop as they make sense of the world around them.

- Each module then contains a sequence of lesson plans. Snap Science is designed around the principle that science should be taught at least once a week throughout the year, as Ofsted recommends.

- The modules are divided into 'core' and 'enrichment' lessons. The 'core' lessons cover all the objectives from the Programme of Study. The 'enrichment' lessons provide extra breadth and depth for the topic.

A SNAP SCIENCE LESSON

Every lesson begins with a **question** – providing a focus for children to explore and think about.

The 'C' or 'E' symbol indicates whether it is a 'core' or 'enrichment' lesson

Key vocabulary highlights important technical and descriptive language that should be used as part of the lesson.

Resources lists all the materials you will need.

The **learning intention** establishes a clear aim for what children should learn over the course of the lesson. Teachers will not always want to share this with children at the beginning of the lesson.

The **Scientific enquiry type** is highlighted where appropriate

The **Explore** section begins the lesson. It introduces the science phenomena, sparking curiosity and enabling children to share their previous understanding and generate questions to investigate.

The **lesson summary** gives a quick overview of the main activity in the lesson, so you can see at a glance what you will be covering.

Each lesson links directly to the **Programmes of Study** and the **Working Scientifically** criteria.

Success criteria for each lesson are written in child speak. They exemplify what successful attainment of the learning intention looks like.

Prompt questions are included throughout as ways of eliciting or developing children's understanding

The **Enquire** section is where children will answer these questions. It is divided into three levels of differentiated challenge to ensure all children can access and master the lesson's learning intention.

MODULE 2

ROCK DETECTIVES

C **LESSON 3: HOW ARE ROCKS USED AROUND OUR SCHOOL?**

Key vocabulary:
rock, sandstone, granite, chalk, limestone, marble, concrete, slate, brick, clay, stone, tile, roof, floor, pavement, wall

Resources:
No extra resources

LESSON SUMMARY:

In this lesson children will identify where and how rocks are used in their local environment. By the end of the lesson children will be able to identify a variety of contexts where rocks have been used and explain why their properties make them particularly suited to the job that they are doing.

National curriculum links:

Compare and group together different kinds of rocks on the basis of their appearance and simple physical properties

Working scientifically links:

Gathering, recording, classifying and presenting data in a variety of ways to help in answering questions

Learning intention:

To recognise where and how rocks are used and explain how their properties make them suitable for their purpose

Scientific enquiry type:

Grouping and classifying

Success criteria:

• I can identify different types of rocks found around the school.

• I can describe how they are being used.

• I can explain why their properties make them useful for this purpose.

EXPLORE:

Explore key questions with children, reviewing their prior learning from the previous two lessons.

Ask: *What rocks do you know of already? What are their properties? Which do you think is the hardest? Which the softest? How might they be used?*

Play 'Rock or not' (Interactive 1) an interactive whiteboard game featuring images of different buildings, structures, street furniture and so on. Encourage the children to discuss the rocks they can see pictured. They then drag and drop images into 'Rock' or 'Not' bins.

ENQUIRE:

Tell the children that they are going to be Rock Detectives, and that their challenge is to find out how rocks are used around school and in places further afield. The challenges are differentiated by the level of detail required in their survey, the support given in recording the investigation and the manner of presentation. The challenges are presented on the Challenge slides to be displayed on the board, or printed out and placed in the centre of the table.

Challenge 1: Children identify and sort rocks in use in the environment around them

Display Rocks in our environment (Slideshow 1) or print out and laminate the photos from the slideshow to give to children. They review, identify, sort and group the images of buildings, structures and objects, some made of rock or rock derivatives, others made of other materials, some of which look like rock, such as plastic garden centre pots.

Ask: *Why was rock used to make some of the objects in the photographs? What types of rocks would be best to use?*

Challenge 2: Children carry out a survey of the rocks found in their local school environment.

Organise children into groups and give them their Rock Checklist (Resource sheet 1). You may wish to group the children and send them to different locations in order to contribute to a more extensive survey of the school and grounds; for example, group 1 corridor, group 2 classrooms, group 3 entrance and school office, group 4 school field, group 5 school gate and roadway. The children name the objects made of rock, suggest what rock they think has been used (using the checklist) and explain why.

Provide them with a table (Resource sheet 2) in which to record their survey findings (either on paper or using iPads) and remind them to act as thorough Rock Detectives, looking carefully and in detail as they complete their task. When they return to the classroom, ask them to review what they have found out.

Prompts are provided throughout the lesson plan to highlight when there is a **related asset** for you to use.

Key information
is provided throughout, providing helpful tips and scientific background knowledge.

Opportunities to link lessons to other subject areas are flagged as appropriate.

White space for teachers to make notes and changes to lesson plans. Snap Science is designed to be supportive not prescriptive!

Key information
Look out for and photograph (if possible) any examples of worn rock surfaces that you see while the children undertake the survey.

Key information
Establish that not all rock-like materials are actually natural rock that has been shaped in some way to fit our purpose. Some rock-like materials have been manufactured in a variety of different ways – like concrete.

LESSON 3: HOW ARE ROCKS USED AROUND OUR SCHOOL?

Ask: *Where did they see rocks being used? What were they used for? Why do you think rock X should not be used for that purpose? What types of rocks did you see most often? Why might this be?*

Challenge 3: Children observe and describe an object made from rock, presenting their findings creatively.

Provide children with a selection of objects made of rock (from a collection in the classroom or from around school). Ask them to choose an object, describe how it has been made and which properties of the rock make it particularly appropriate for the job that it does. Children then communicate their findings creatively, through multi-media where possible. They need to consider (and feed back on) a number of 'what do you think would happen if' questions, such as, 'what would happen if your object was left in the rain?' 'What would happen if it was exposed to very hot or cold temperatures?' 'What would happen if it was dropped?

REFLECT AND REVIEW:

Ask children to consider the evidence that they collected as part of Challenge 2.

Ask: *Which rocks were seen most frequently? How were rocks being used? Where were rocks most often used? Were there any uses for rocks that surprised you?*

Explore children's understanding of 'tricky' rock-based materials, such as concrete, manufactured roof tiles, breeze blocks.

Ask: *What are they made of? How are they made? What makes them useful? Who invented them and how?*

EVIDENCE OF LEARNING:

Watch and listen to the children as they discuss what they have found out. Can they name a variety of rocks they have seen used around the school? Can they describe how different rocks have been used? Can they identify where rocks have been changed in some way to make them useful? (For example, a rock derivative, which is made up of reconstituted rock fragments made into a new solid rock.) Can they explain what it is about a particular rock that makes it suitable for its purpose?

CROSS-CURRICULAR OPPORTUNITIES:

IT – Multi media presentation

The **Reflect and Review** section provides the opportunity for children to consolidate the key learning from the lesson.

Evidence of Learning provides assessment guidance for teachers, indicating the kinds of things children might say, write, draw or do to demonstrate they have achieved or exceeded the learning intention.

THE ONLINE TOOLKIT

- You can access all of the lesson sequences, lesson plans and related assets online via the **Collins Connect platform**. There is a discrete area for each year group, with support for planning, teaching and tracking progression.

- Simply select your year group using the tabs along the top. Using the drop down lists, choose the module you would like to teach and then a lesson from the sequence. Click on the lesson to reveal the lesson plan and related assets.

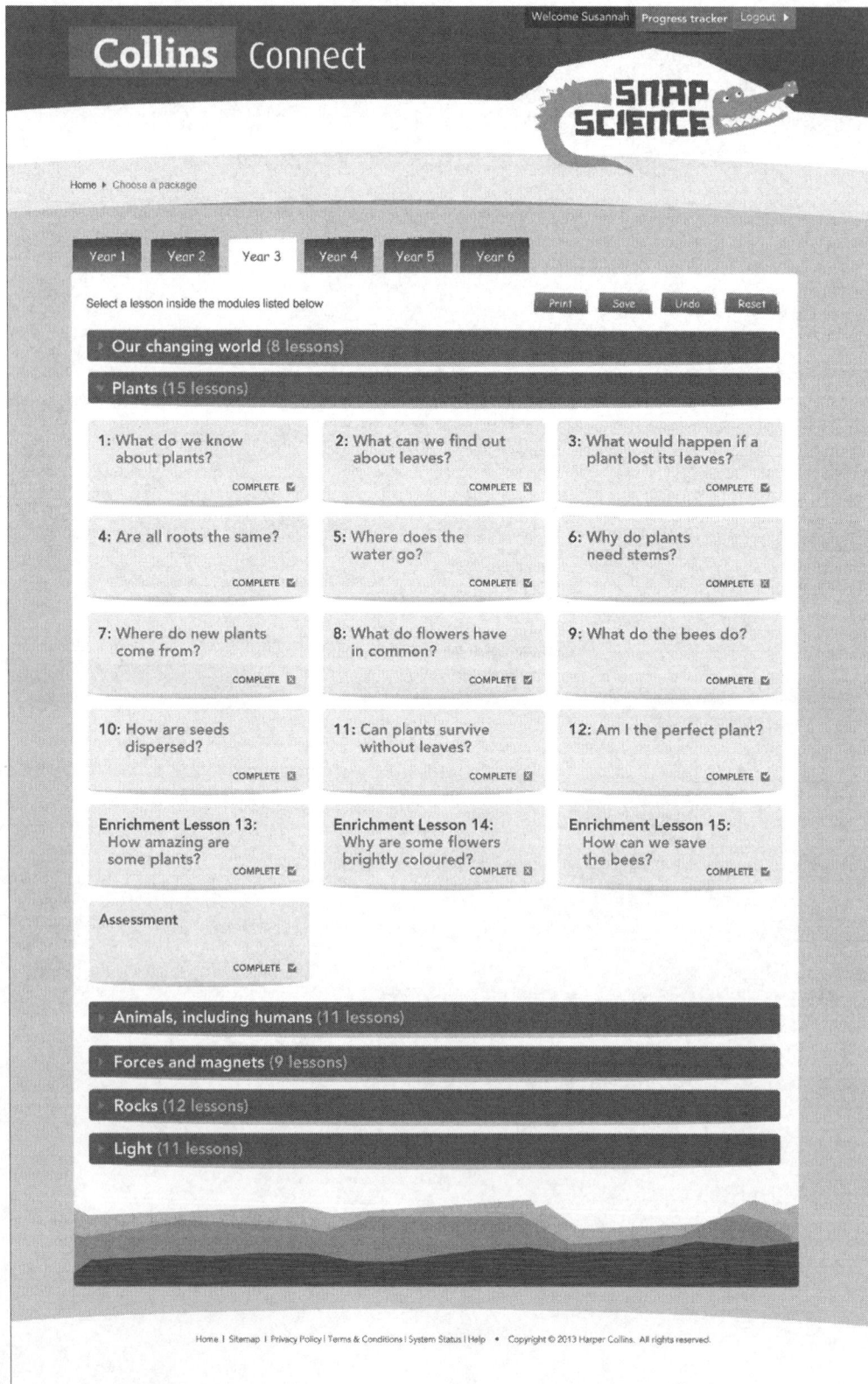

- **The lesson plan is provided in an editable format**, so you can adapt and customise the content for your class.
- A **range of digital assets** are provided to help make your science lessons rich, lively and engaging.
- All of the assets for the lesson are clearly labelled, with the **asset type highlighted** so you can easily navigate to each one.

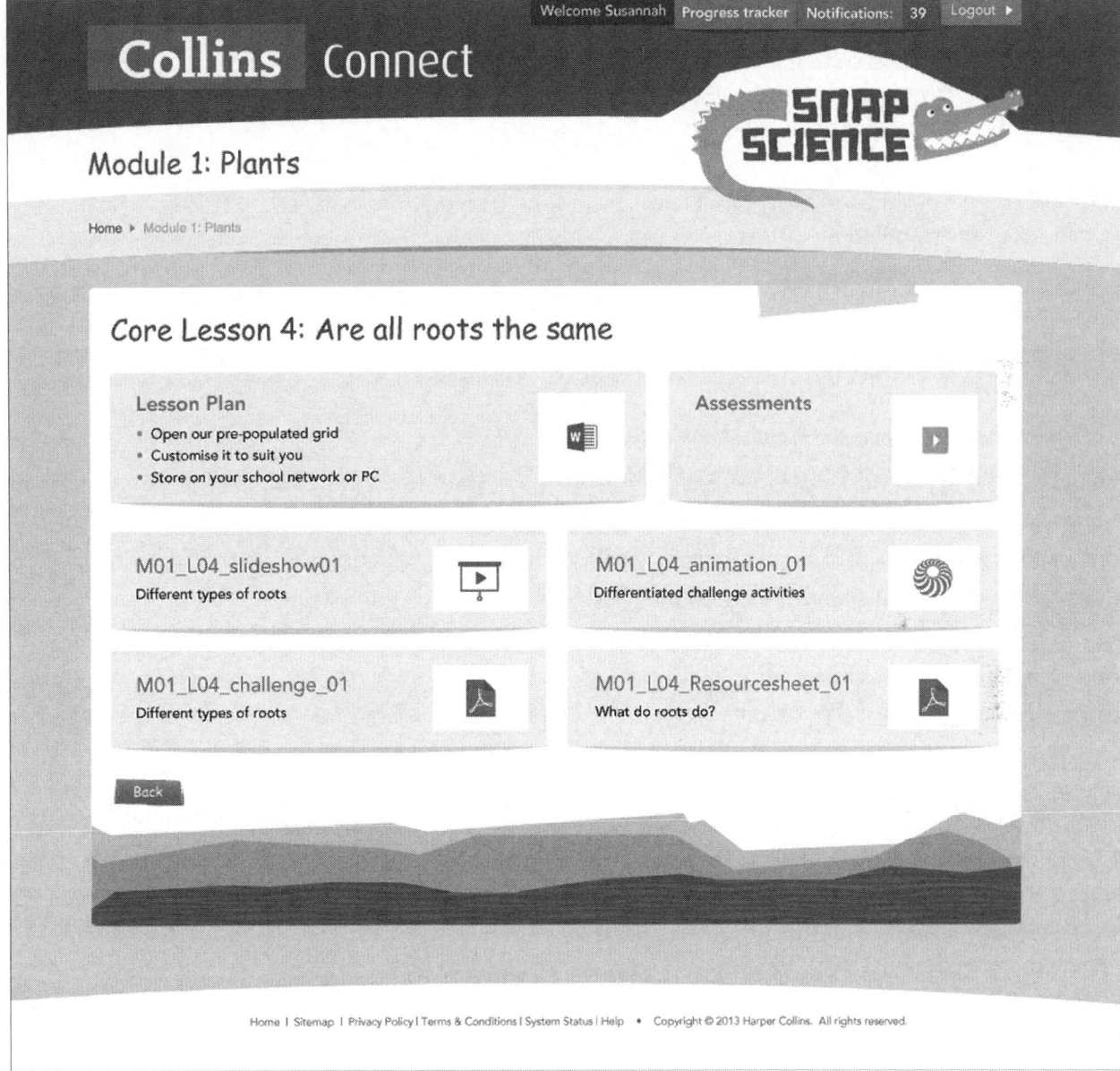

PROGRESSIVE SUCCESS CRITERIA FOR DATA COLLECTION AND ANALYSIS SKILLS

In every lesson in Snap Science success criteria are used to break down the learning intention to help children to identify what the steps to success are. Often success criteria refer to the specific skills of collecting, presenting and interpreting data, for example *I can record my observations in a table*.

For more detailed understanding of what success looks like in these data collection and analysis skills, teachers and children can use the following sets of progressive success criteria. They should be used formatively to ensure that children are motivated to reach goals that are specific, within reach and offer some degree of challenge; also summatively by children and teachers to assess the level of success and therefore plan the next steps.

Measuring

- I can take non-standard measures
- I can choose equipment to take measurements
- I can use standard measures
- I can start measuring from zero
- I can read the numbers on a scale
- I can estimate between the numbers on a scale
- I can convert between different units of metric measure e.g. m/km
- I can decide how accurate to make my measurements
- I can choose equipment with an appropriate data range
- I can recognise that a series of measurements should be made
- I can select suitable ranges and intervals
- I can make sure I collect enough data
- I can repeat readings to check my results
- I can use compound units for speed

Collecting data

- I can record my results in a tally chart
- I can construct a tally chart
  - I can draw a simple chart with three columns
  - I can put the different things I am counting in separate rows
  - I can count in 5's making tally marks
  - I can add up the frequency in the final column
- I can record my results in a table
- I can construct a table
  - I can draw a simple table using two lines
  - I can write what I am changing in the first column
  - I can write what I am measuring in the second column
  - I can include the units
- I can add extra columns to my table for repeat readings
- I can calculate the average
- I can add a column for the average reading

Presenting results

- I can present my results as a pictogram
- I can construct a pictogram
  - I can decide what picture to use
  - I can include a key
- I can present my results as a bar graph
- I can construct a bar graph
  - I can draw the axes using a ruler
  - I can label the axes
  - I can plot what I have changed (the independent variable) on the horizontal x axis and what I have measured in order to collect the results (the dependent variable) on the vertical y axis
  - I can choose an appropriate scale for the y axis
  - I can choose a scale that goes up to a suitable number
  - I can choose a scale that increases by regular intervals
  - I can draw bars of the correct height
- I can construct a bar line graph
- I can present my results as a line graph
- I can construct a line graph
  - I can draw the axes using a ruler
  - I can label the axes
  - I can plot what I have changed (the independent variable (what is changed) goes on the horizontal x axis and what I have measured in order to collect the results (the dependent variable) on the vertical y axis
  - I can choose an appropriate scale for the axes
  - I can choose a scale that goes up to a suitable number
  - I can choose a scale that increases by regular intervals
  - I can plot points accurately
  - I can join the points
- I can choose which type of graph to use

Interpreting data

- I can answer questions about one value in a pictogram by counting the pictures
- I can answer questions about one value in a block graph or line graph by using the numbers on the y axis
- I can answer comparative questions about two or more values in a pictogram or block graph
- I can ask questions using the points on a line graph
- I can answer questions using the line on a line graph
- I can read between the points on a line graph
- I can use the line graph to make predictions outside the data collected
- I can use a pictogram, bar graph or bar line graph to make statements about quantity/ measurements or frequency
- I can use a line graph to make statements about patterns between two sets of continuous data

SUCCESS CRITERIA FOR DIFFERENT APPROACHES TO SCIENCE ENQUIRY

These success criteria can be used by teachers and children to plan different aspects of scientific enquiry

Comparative test – comparing one thing with another to investigate a difference between them
- I can recognise when to use a comparative test to answer a question
- I can identify properties or features to compare
- I can choose how to observe or measure the feature I am comparing
- I can decide what equipment to use
- I can decide how to collect and present data
- I can write a clear question for my comparative test
- I can explain how I made my results reliable

Fair test – controlled testing to investigate the impact of changing one variable on another
- I can recognise when to use a fair test to answer a question
- I can identify (independent) variables to change
- I can identify (dependent) variables to observe or measure
- I can choose the (independent) variable to change
- I can choose the (dependent) variable to measure
- I can identify the variable to keep the same (control)
- I can identify which variables cannot be controlled
- I can decide what equipment to use
- I can decide how to collect and present the data
- I can write a clear question for my fair test
- I can explain how I will make the test fair

Noticing patterns – when children make observations or carry out surveys of variables that cannot be easily controlled to investigate relationships between two sets of data
- I can recognise when to look for patterns to answer a question
- I can identify patterns I could investigate
- I can choose two variables to observe or measure
- I can decide what equipment to use
- I can decide on sample size
- I can decide how to collect and present my data
- I can write a clear question for my pattern seeing investigation

Observing changes over time – when children observe or measure how one variable changes over time to investigate how change takes place
- I can recognise when to make observations over time to answer a question
- I can identify variables to observe or measure
- I can decide how often to make observations or take measurements
- I can decide how long to continue to make observations or take measurements
- I can decide what equipment to use
- I can decide how to collect and present my data
- I can write a clear question for my observing over time investigation

Grouping and classifying – when children identify materials and living things and make observations or carry out tests to identify similarities and differences and organise them into groups

- I can recognise when to identify and classify objects, materials or living things to answer a question
- I can identify observable differences and similarities to observe or measure
- I can decide what equipment to use
- I can decide how to collect and present data
- I can identify behavioural differences and similarities to observe or measure
- I can identify simple tests to carry out to identify differences and similarities
- I can write a clear question for my identifying and classifying investigation

Finding things out using secondary sources of information – when children answer questions using data or information that they have not collected first hand

- I can recognise when to use secondary data to answer a question
- I can suggest how to find things out
- I can decide what information sources to use
- I can decide how to collect and present data
- I can write a clear question for my research using secondary data investigation

PROGRESSION CHARTS

The National Curriculum Programme of Study for Science describes a sequence of knowledge and concepts, processes and methods. This sequence of knowledge and concepts is arranged as progressive blocks of key ideas in biology, chemistry and physics, alongside a progression in the skills of working scientifically.

The conceptual ideas in Biology, Chemistry and Physics build on each other and children need to develop a strong understanding of each set of ideas in order for the next set to make sense and for them to make progress. The Programme of Study is set out year by year for Key stages 1 and 2 but each science topic is not covered in every year. It is therefore important that teachers and children know where each block of ideas fits into the overall sequence.

In the Snap Science Progression Charts the key ideas within Biology, Chemistry and Physics in the National Curriculum are arranged to show how they are related to each other and how one idea builds on another. The National Curriculum statements have been edited into key ideas statements. The source of each key idea is identified by the year group and the Programme of Study topic heading. Some additional statements have been added to make important links between ideas.

Working Scientifically is taught throughout KS1 and 2, embedded within the content of Biology, Chemistry and Physics. The National Curriculum Programme of Study for Working Scientifically outlines the practical scientific methods, processes and skills that children must be taught to use, divided into three two-year blocks. In every lesson in Snap Science children will use their developing science enquiry skills to answer scientific questions. The Snap Science Progression Chart for Working Scientifically exemplifies the progression in these skills in the key areas of raising questions and planning, collecting and presenting data, drawing and evaluating conclusions.

This progression underpins the sequence of teaching and learning in each Snap Science module and between year groups.

BIOLOGY: progression of ideas through KS1 and 2

LIFE PROCESSES

STRUCTURE AND FUNCTION

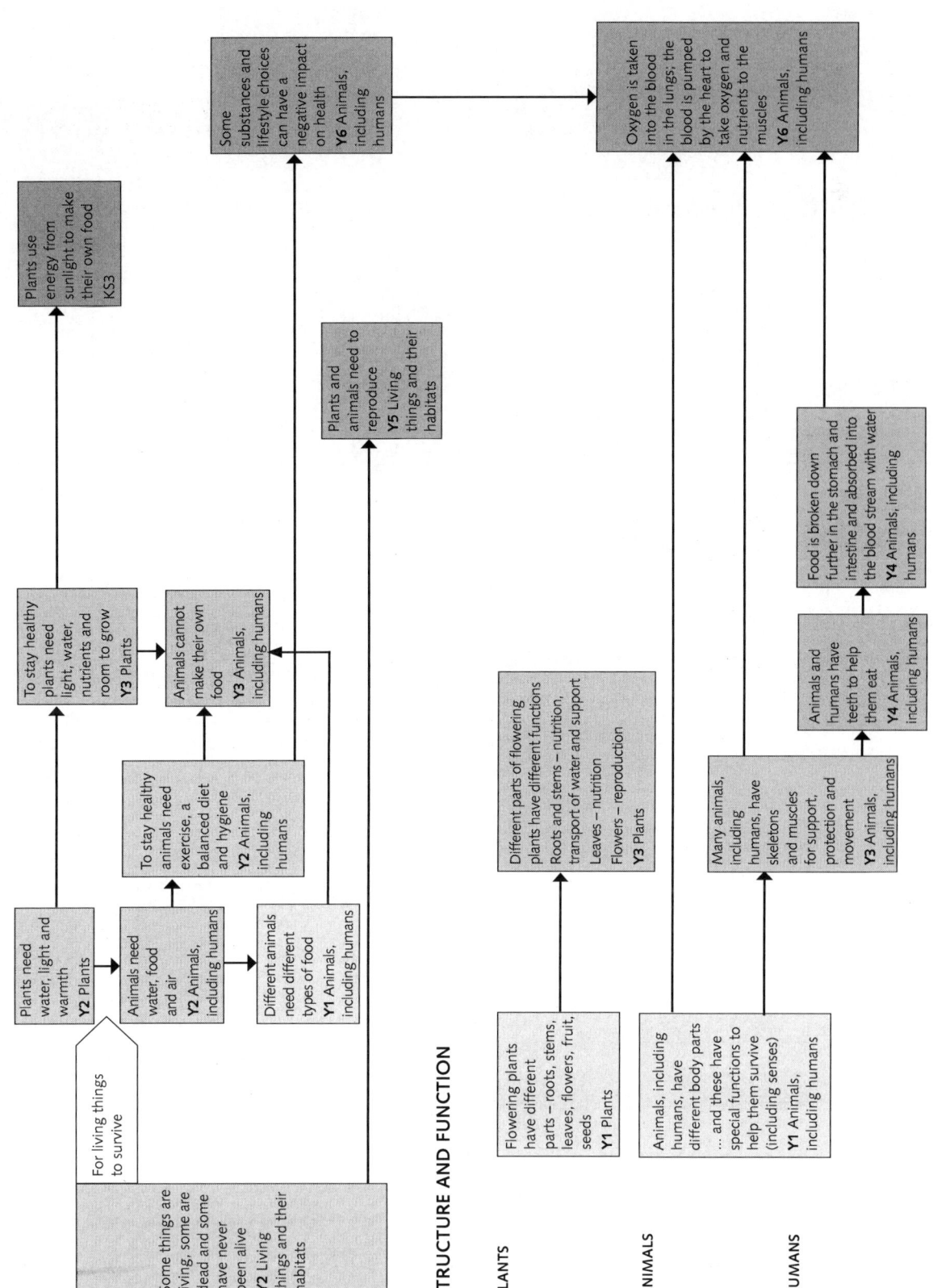

PLANTS

ANIMALS

HUMANS

CLASSIFICATION

Identifying and classifying increasing range from the familiar to the unfamiliar

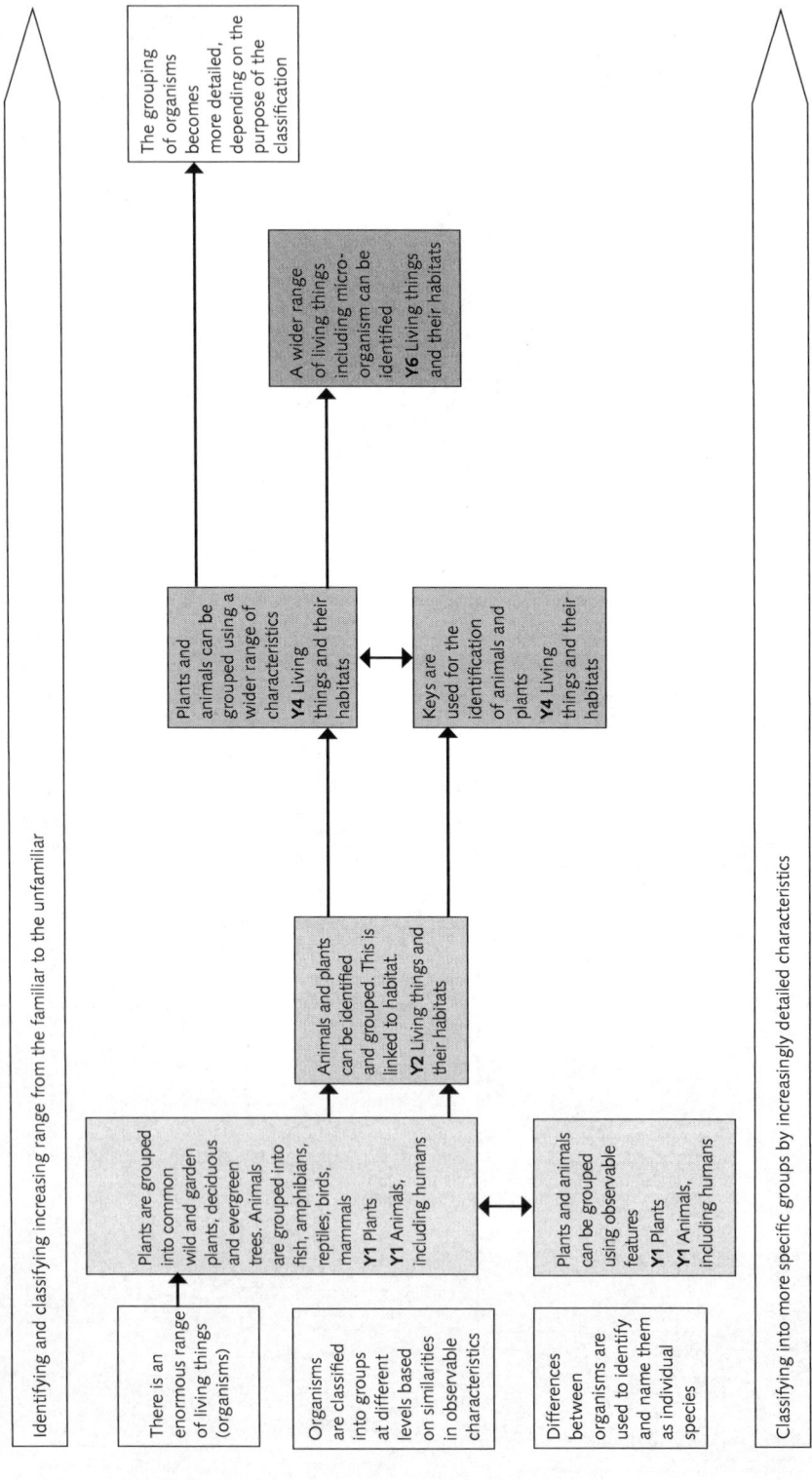

There is an enormous range of living things (organisms)

Organisms are classified into groups at different levels based on similarities in observable characteristics

Differences between organisms are used to identify and name them as individual species

Plants are grouped into common wild and garden plants, deciduous and evergreen trees. Animals are grouped into fish, amphibians, reptiles, birds, mammals
Y1 Plants
Y1 Animals, including humans

Plants and animals can be grouped using observable features
Y1 Plants
Y1 Animals, including humans

Animals and plants can be identified and grouped. This is linked to habitat.
Y2 Living things and their habitats

Plants and animals can be grouped using a wider range of characteristics
Y4 Living things and their habitats

Keys are used for the identification of animals and plants
Y4 Living things and their habitats

A wider range of living things including micro-organism can be identified
Y6 Living things and their habitats

The grouping of organisms becomes more detailed, depending on the purpose of the classification

Classifying into more specific groups by increasingly detailed characteristics

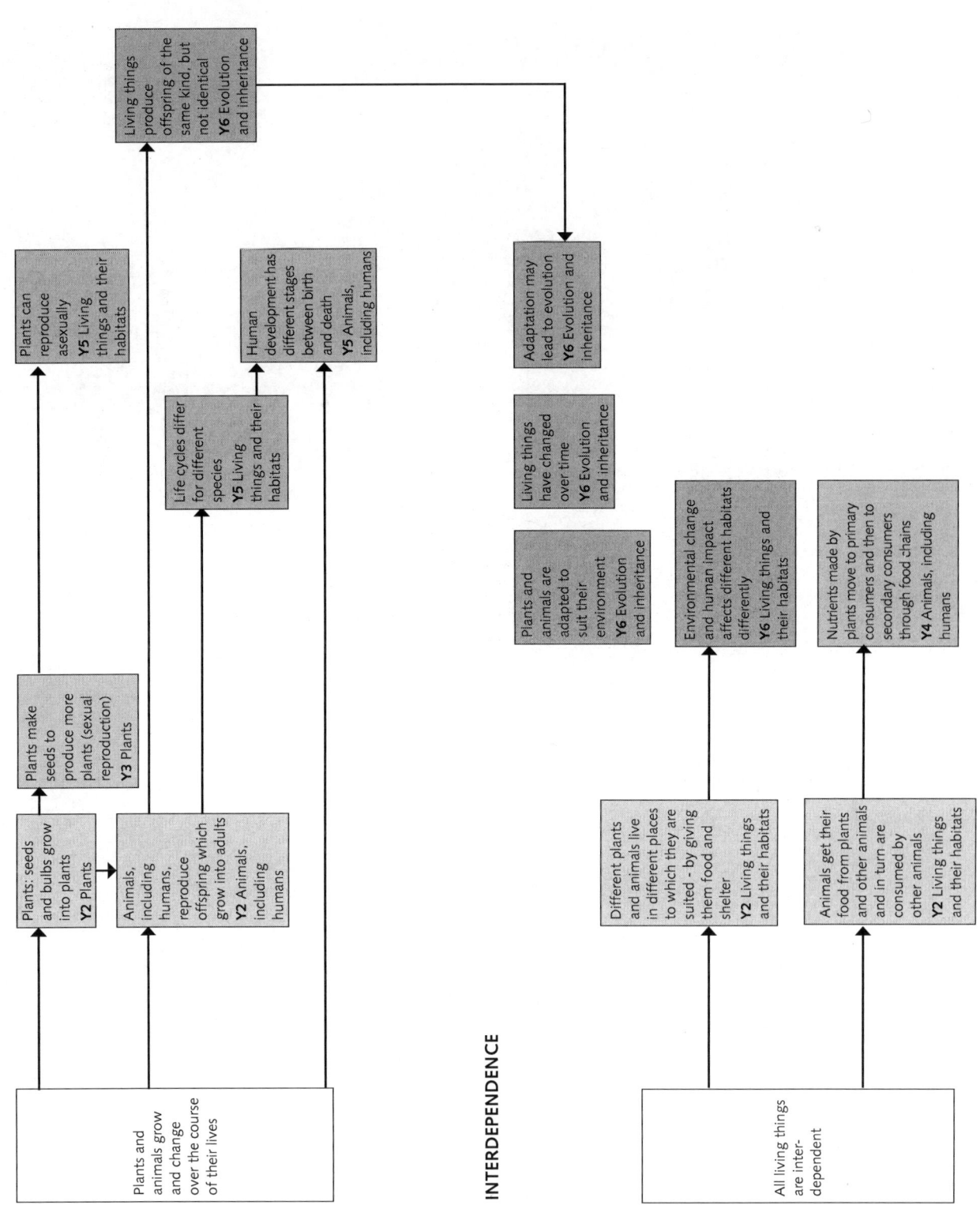

LIFE CYCLES

Plants and animals grow and change over the course of their lives

Plants: seeds and bulbs grow into plants
Y2 Plants

Plants make seeds to produce more plants (sexual reproduction)
Y3 Plants

Plants can reproduce asexually
Y5 Living things and their habitats

Animals, including humans, reproduce offspring which grow into adults
Y2 Animals, including humans

Life cycles differ for different species
Y5 Living things and their habitats

Human development has different stages between birth and death
Y5 Animals, including humans

Living things produce offspring of the same kind, but not identical
Y6 Evolution and inheritance

INTERDEPENDENCE

All living things are inter-dependent

Different plants and animals live in different places to which they are suited - by giving them food and shelter
Y2 Living things and their habitats

Animals get their food from plants and other animals and in turn are consumed by other animals
Y2 Living things and their habitats

Environmental change and human impact affects different habitats differently
Y6 Living things and their habitats

Nutrients made by plants move to primary consumers and then to secondary consumers through food chains
Y4 Animals, including humans

Plants and animals are adapted to suit their environment
Y6 Evolution and inheritance

Living things have changed over time
Y6 Evolution and inheritance

Adaptation may lead to evolution
Y6 Evolution and inheritance

CHEMISTRY: progression of ideas through KS1 and 2

MATERIALS

DESCRIBING AND USING MATERIALS

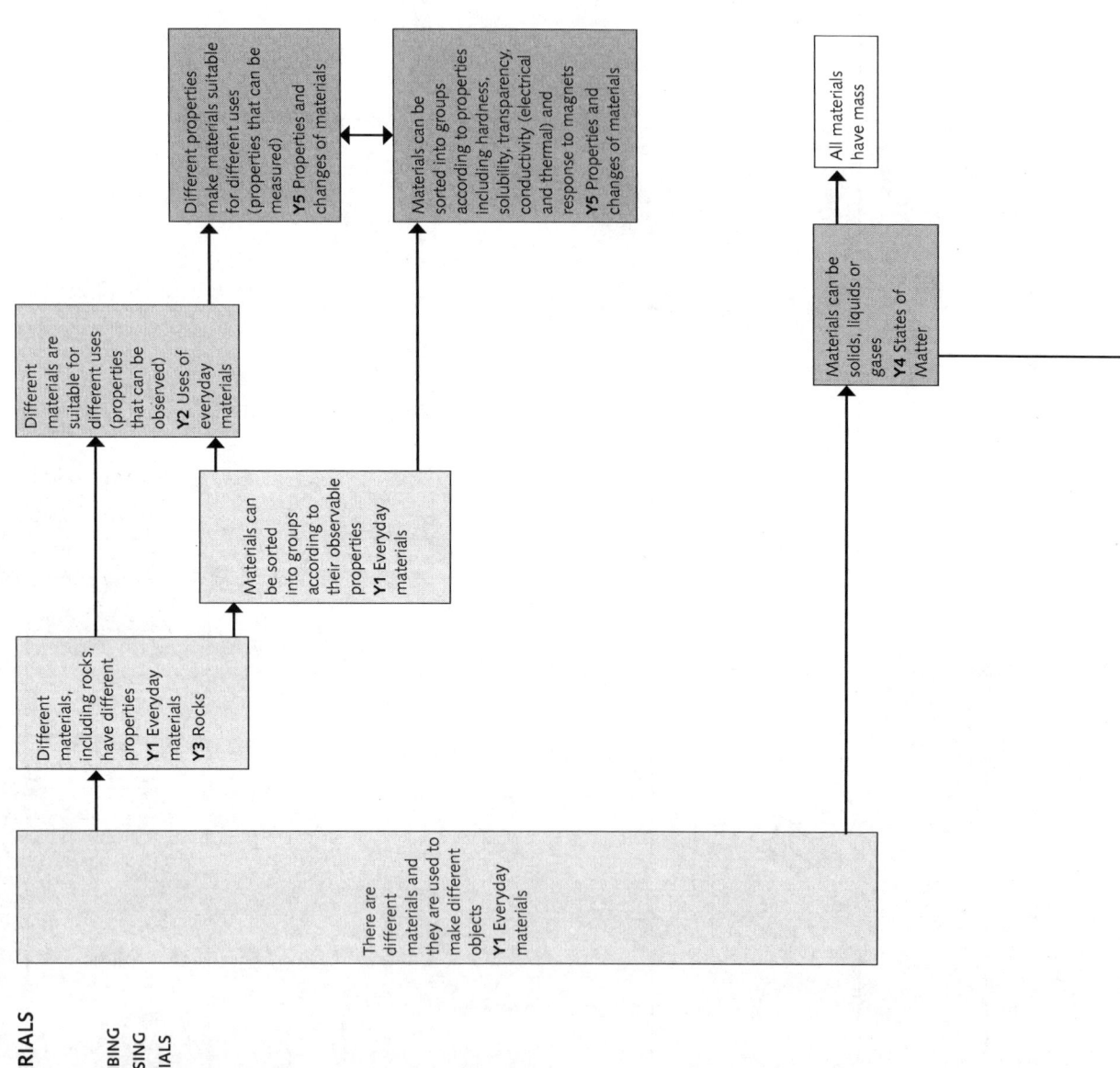

Different materials, including rocks, have different properties
Y1 Everyday materials
Y3 Rocks

Different materials are suitable for different uses (properties that can be observed)
Y2 Uses of everyday materials

Different properties make materials suitable for different uses (properties that can be measured)
Y5 Properties and changes of materials

Materials can be sorted into groups according to their observable properties
Y1 Everyday materials

Materials can be sorted into groups according to properties including hardness, solubility, transparency, conductivity (electrical and thermal) and response to magnets
Y5 Properties and changes of materials

There are different materials and they are used to make different objects
Y1 Everyday materials

Materials can be solids, liquids or gases
Y4 States of Matter

All materials have mass

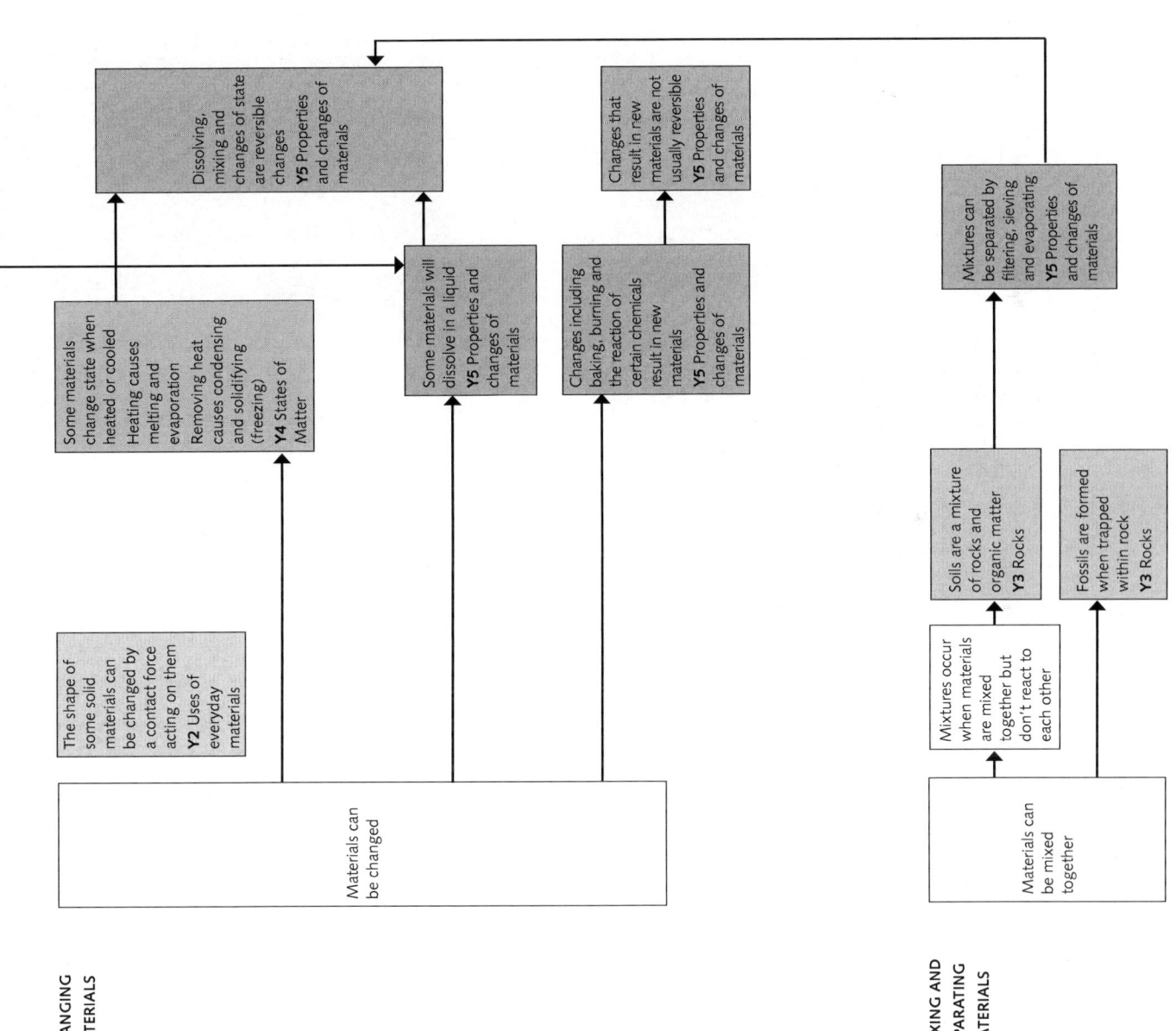

CHANGING MATERIALS

MIXING AND SEPARATING MATERIALS

PHYSICS: progression of ideas through KS1 and 2

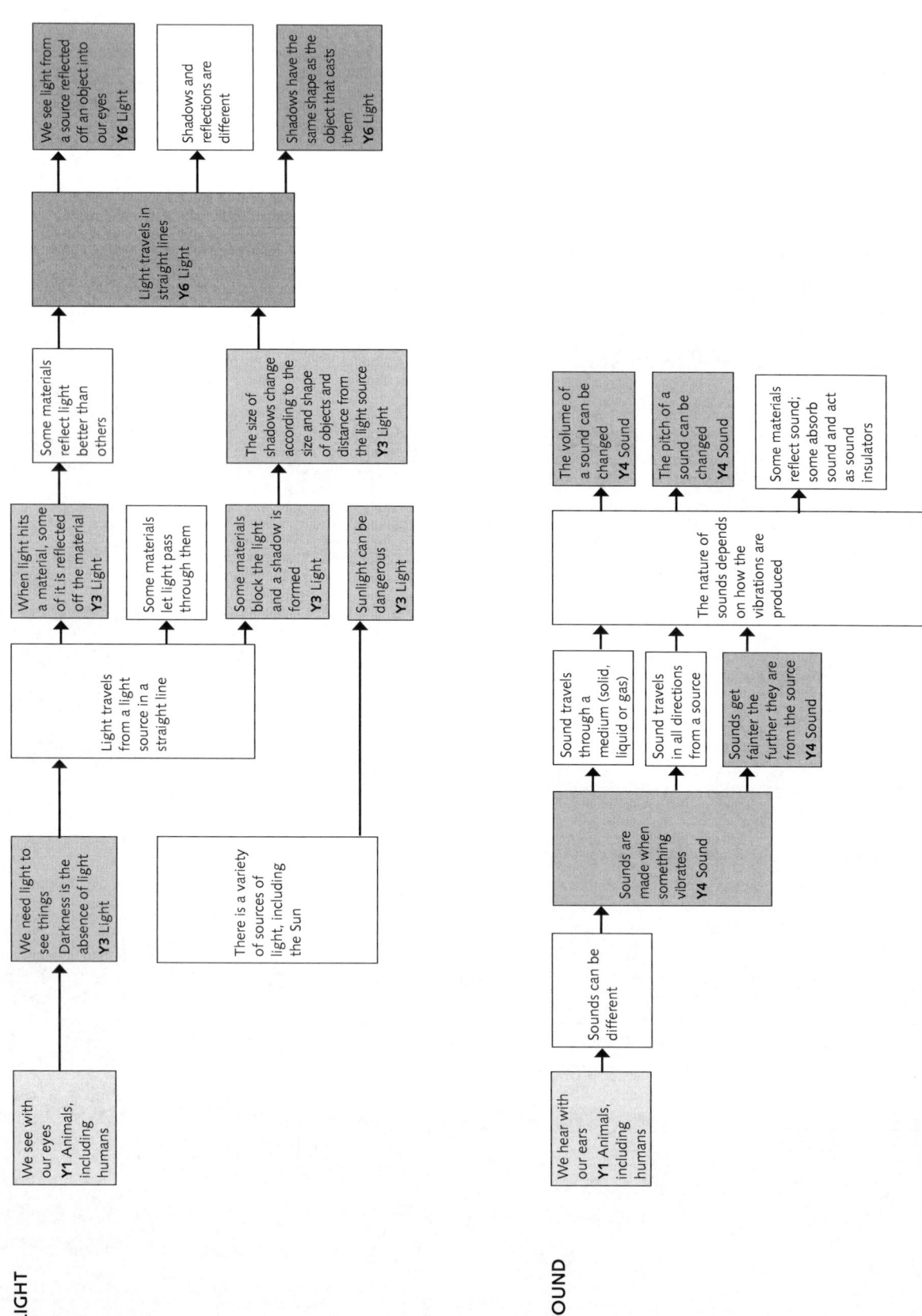

LIGHT

We see with our eyes
Y1 Animals, including humans

We need light to see things
Darkness is the absence of light
Y3 Light

When light hits a material, some of it is reflected off the material
Y3 Light

Some materials reflect light better than others

We see light from a source reflected off an object into our eyes
Y6 Light

Light travels in straight lines
Y6 Light

Shadows and reflections are different

Shadows have the same shape as the object that casts them
Y6 Light

Light travels from a light source in a straight line

Some materials let light pass through them

There is a variety of sources of light, including the Sun

Some materials block the light and a shadow is formed
Y3 Light

Sunlight can be dangerous
Y3 Light

The size of shadows change according to the size and shape of objects and distance from the light source
Y3 Light

SOUND

We hear with our ears
Y1 Animals, including humans

Sounds can be different

Sounds are made when something vibrates
Y4 Sound

Sound travels through a medium (solid, liquid or gas)

Sound travels in all directions from a source

Sounds get fainter the further they are from the source
Y4 Sound

The nature of sounds depends on how the vibrations are produced

The volume of a sound can be changed
Y4 Sound

The pitch of a sound can be changed
Y4 Sound

Some materials reflect sound; some absorb sound and act as sound insulators

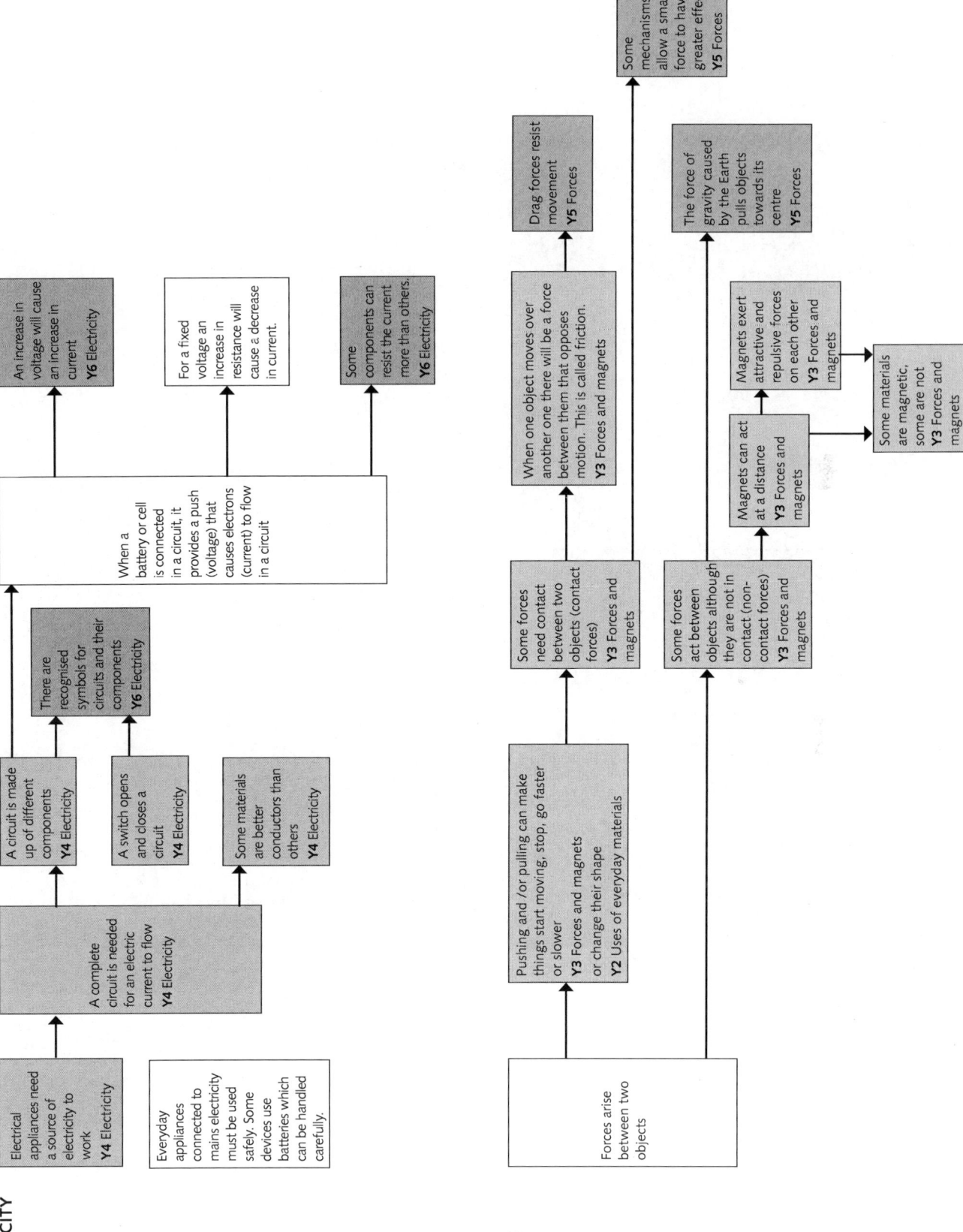

ELECTRICITY

Electrical appliances need a source of electricity to work **Y4** Electricity

Everyday appliances connected to mains electricity must be used safely. Some devices use batteries which can be handled carefully.

A complete circuit is needed for an electric current to flow **Y4** Electricity

A circuit is made up of different components **Y4** Electricity

A switch opens and closes a circuit **Y4** Electricity

Some materials are better conductors than others **Y4** Electricity

There are recognised symbols for circuits and their components **Y6** Electricity

When a battery or cell is connected in a circuit, it provides a push (voltage) that causes electrons (current) to flow in a circuit

An increase in voltage will cause an increase in current **Y6** Electricity

For a fixed voltage an increase in resistance will cause a decrease in current.

Some components can resist the current more than others. **Y6** Electricity

FORCES

Forces arise between two objects

Pushing and /or pulling can make things start moving, stop, go faster or slower **Y3** Forces and magnets or change their shape **Y2** Uses of everyday materials

Some forces need contact between two objects (contact forces) **Y3** Forces and magnets

Some forces act between objects although they are not in contact (non-contact forces) **Y3** Forces and magnets

When one object moves over another one there will be a force between them that opposes motion. This is called friction. **Y3** Forces and magnets

Magnets can act at a distance **Y3** Forces and magnets

Magnets exert attractive and repulsive forces on each other **Y3** Forces and magnets

Some materials are magnetic, some are not **Y3** Forces and magnets

Drag forces resist movement **Y5** Forces

The force of gravity caused by the Earth pulls objects towards its centre **Y5** Forces

Some mechanisms allow a smaller force to have a greater effect **Y5** Forces

EARTH IN SPACE

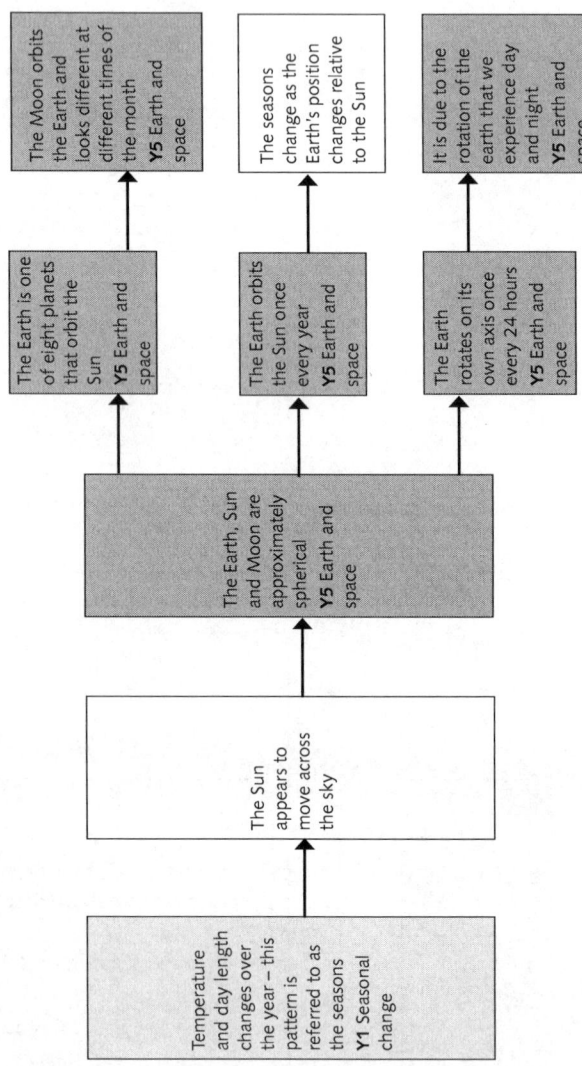

Temperature and day length changes over the year – this pattern is referred to as the seasons **Y1** Seasonal change

The Sun appears to move across the sky

The Earth, Sun and Moon are approximately spherical **Y5** Earth and space

The Earth is one of eight planets that orbit the Sun **Y5** Earth and space

The Moon orbits the Earth and looks different at different times of the month **Y5** Earth and space

The Earth orbits the Sun once every year **Y5** Earth and space

The seasons change as the Earth's position changes relative to the Sun

The Earth rotates on its own axis once every 24 hours **Y5** Earth and space

It is due to the rotation of the earth that we experience day and night **Y5** Earth and space

WORKING SCIENTIFICALLY

Developing independence and autonomy in raising questions, planning and carrying out investigations

Approaches to enquiry

Children should be helped to develop their understanding of scientific ideas by using different types of scientific enquiry to answer their own questions, including:

- observing changes over a period of time
- noticing patterns
- grouping and classifying things
- carrying out simple comparative tests
- finding things out using secondary sources of information

Children should ask their own questions about what they observe and make some decisions about which types of scientific enquiry are likely to be the best ways of answering them including:

- observing changes over time
- noticing patterns
- grouping and classifying things
- carrying out simple fair tests
- finding things out using secondary sources of information

Children should select the most appropriate ways to answer science questions using different types of scientific enquiry, including:

- observing changes over different periods of time
- noticing patterns
- grouping and classifying things
- carrying out fair tests
- finding things out using a wide range of secondary sources of information

Asking questions

Ask simple questions

- Begin to shape questions using different question stems
- Ask questions about how and why objects, materials and living things:
  o change
  o are similar or different to each other
  o connect with each other
  o are made or work
- Suggest questions to investigate

Ask relevant questions

- Recognise questions that can be investigated scientifically and those that cannot
- Ask a clear scientific question
- Recognise when questions can be answered by first hand or second sources of evidence

Use results to raise further questions

- Independently ask questions and offer ideas for scientific enquiry

Use test results to make predictions to set up further comparative and fair tests

Planning

Recognise that questions can be answered in different ways

- With support:
  o Suggest how to find things out
  o Identify changes to observe and measure
  o Identify patterns to observe and measure
  o Identify variables to change and measure
  o Identify sorting criteria
  o Suggest how to take measurements
  o Suggest next steps or a sequence of steps in a plan

Use different types of scientific enquiries to answer them

- Identify different ways to answer a question
- Choose the most appropriate method

Set up simple practical enquiries, comparative and fair tests

- Decide what observations to make, how often and what equipment to use
- Decide what measurements to take, how long to make them for and whether to repeat them
- Decide what sorting or classification criteria to use
- Recognise when a simple fair test is necessary
- With help, decide what variables to change and measure

Plan different types of scientific enquiries to answer questions

- Explain why an enquiry method is the most appropriate to answer a question
- Plan systematic collection of data and which equipment to use
- Plan collection of sufficient data
- Recognise when research using secondary sources will answer questions
- Decide which sources of information to use to answer questions

Recognise and control variables where necessary

- Recognise when variables need to be controlled and why
- Recognise when variables cannot be controlled and a pattern seeking enquiry is appropriate
- Identify which variables have the greatest effect on the result

Becoming more systematic and accurate in collecting, recording and presenting data

Collecting data

Observe closely, using simple equipment

- Choose and use appropriate simple equipment to make observations
- Use non-standard units to collect observations

performing simple tests

- Choose and use appropriate simple equipment with increasing accuracy to collect comparative data
- Use non-standard units to collect data

identifying and classifying

- Sort objects by observable and behavioural features
- Make comparisons between simple features

gathering data to help in answering questions

- Gather data to answer questions from a variety of sources including talking to people, simple books and electronic media, first hand observation and practical activity

Make systematic and careful observations where appropriate, take accurate measurements using standard units, using a range of equipment, including thermometers and data loggers

- Use a range of equipment including data loggers to collect data using standard measures
- With support take accurate measurements on measuring equipment, recognising when to repeat them
- Carry out simple tests to sort and classify materials according to properties or behaviour

Gather data in a variety of ways to help in answering questions

- Gather data to answer questions from a variety of sources including using textbooks, simple keys, electronic media, first hand observation, practical activity and data collected by others

Take measurements, using a range of scientific equipment with increasing accuracy and precision

- Use a range of equipment accurately without support to collect observations and measurements
- Repeat sets of observations or measurements, where appropriate, selecting suitable ranges and intervals
- Use a series of tests to sort and classify materials
- Use relevant information and data from a range of secondary sources to answer questions

Presenting data

Record data to help in answering questions

- Talk about what has been found out and how
- Record observations in word and pictures
- Record observations and test results in simple prepared pictograms, tables, tally charts, bar charts and maps including ICT formats
- Record sorting in sorting circles or tables

Record data in a variety of ways to help in answering questions

- Make notes
- Record data in tables and bar charts
- Use graphs produced by data loggers

Classify in a variety of ways to help in answering questions

- Use Carroll diagrams, and Venn diagrams to classify
- Use and make simple keys to identify and classify

Present data in a variety of ways to help in answering questions

- Drawings, labelled diagrams
- Bar charts, bar line graphs, simple scatter graphs and tables using ICT where appropriate

Record data and results of increasing complexity using scientific diagrams and labels, classification keys, tables and bar and line graphs and models

- Decide how to record data accurately and appropriately
- Use appropriate scientific language in oral and written presentations
- Make keys and branching databases with 4 or more items
- Use more than one source of scientific evidence to identify and classify things
- Present data in line graphs, scatter graphs and frequency charts

Increasingly using scientific knowledge and understanding in conclusions and explanations

Concluding

Use their observations and ideas to suggest answers to questions

- Use simple scientific language to talk about observation or findings
- Use results to answer the investigation question
- Identify simple changes
- Sequence changes
- Say whether the change was expected
- Identify similarities and differences
- Make simple comparisons
- Make links between two sets of observations
- Identify simple patterns and talk about them
- Say whether the pattern was expected
- Identify simple causal relationships
- Say if the relationship was expected

Report on findings from enquiries, including oral and written explanations, displays or presentations of results and conclusions

- Draw simple conclusions about changes observed and link these to scientific ideas
- Refer to a table or graph when reporting findings
- Begin to use and interpret graphs produced by data loggers
- Draw a simple conclusion about similarities and differences identified and link these to scientific ideas
- Draw conclusions about simple patterns between two sets of data
- Draw simple causal conclusions from fair tests
- Draw conclusions from data from different secondary sources

Identify differences, similarities or changes related to simple scientific ideas and processes

- Make links between:
  - o observed changes
  - o similarities and differences
  - o simple patterns between two sets of data
  - o simple causal relationships
  - o data from secondary sources
- and simple scientific ideas and processes

Use straightforward scientific evidence to answer questions or to support their findings

Refer to evidence from practical tests and observations or from secondary data sources when answering questions or explaining findings

- Use simple scientific language in a range of oral and written presentations suitable for different audiences to present findings

Report and present findings from enquiries, including conclusions, causal relationships and explanations of results in written forms such as displays and other presentations

- Use scientific evidence to answer questions or support findings
- Draw valid conclusions about changes, similarities and differences, and causal relationships from data collected
- Draw valid conclusions that utilise more than one piece of supporting evidence
- Use scientific knowledge to explain findings
- Use simple models to help describe scientific ideas
- Explain differences in repeated observations or measurements, identifying reasons for any anomalies noticed

Communicate findings in written form, displays, multi-media and other forms of presentation using scientific language

Evaluating

- Say whether data was useful
- Say whether an information source was useful

Give an opinion about some further information

Use results to draw simple conclusions, make predictions for new values, suggest improvements, and raise further questions

- Make predictions for new values within or beyond the collected data collected
- Identify new questions arising from the data
- Find ways of improving enquiries

Identify scientific evidence that has been used to support or refute ideas or arguments

- Begin to separate opinion from fact
- Use scientific evidence to justify ideas
- Talk about how scientific ideas have developed over time

Identify when further tests and observations might be needed

Evaluate the effectiveness of their working methods, making practical suggestions for improving them

MODULE OVERVIEW CHART

Module number and name	Lesson number and name	National curriculum links	Working scientifically links	Scientific enquiry type	Lesson summary
Year 5 Module Our Changing World	1: What signs of plant reproduction can we observe around our school?	Describe the life process of reproduction in some plants and animals	Recording data and results of increasing complexity using scientific diagrams and labels, classification keys, tables, scatter graphs and bar and line graphs	Observing changes over different periods of time	During these lessons children identify a variety of plants to observe, visit them regularly throughout the year and look for evidence of plant reproduction, for example, flowers, seed heads, berries and fruits on plants.
	2: How can we grow more plants, without using seeds?	Describe the life process of reproduction in some plants and animals	Identifying scientific evidence that has been used to support or refute ideas or arguments	Observing changes over different periods of time	During these lessons children explore practically some of the methods for growing new plants.
	3: Which plants are best to plant in our growing space? How can we ensure that produce is ready at the right time?	Describe the life process of reproduction in some plants and animals	Planning different types of scientific enquiries to answer questions, including recognising and controlling variables where necessary	Observing changes over different periods of time	During these lessons children apply their knowledge and understanding of plant life cycles as they plan for and grow plants across the year, ready for summer term 'produce sales' or similar events.
	4: How can we ensure that plants in our growing space yield as many crops as possible?	Describe the life process of reproduction in some plants and animals	Planning different types of scientific enquiries to answer questions, including recognising and controlling variables where necessary	Carrying out comparative and fair tests	During these lessons children set up and carry out an investigation to test a variable that may affect crop production.
Year 5 Module 1 Circle of Life	1: What is a life cycle?	Explain the differences in the life cycles of a mammal, an amphibian, an insect and a bird	Reporting and presenting findings from enquiries, including conclusions, causal relationships and explanations of and degree of trust in results, in oral and written forms such as displays and other presentations	Using a wide range of secondary sources of information	In this lesson children are introduced to the life cycles of four significant types of animals: mammals, amphibians, insects and birds. They compare and contrast different animal life cycles, identifying common features and differences.
	2: What do we know about the life cycles of mammals?	Explain the differences in the life cycles of a mammal, an amphibian, an insect and a bird	Reporting and presenting findings from enquiries, including conclusions, causal relationships and explanations of and degree of trust in results, in oral and written forms such as displays and other presentations	Using a wide range of secondary sources of information	In this lesson children deepen their knowledge about the group of animals called mammals. They find out about the life cycles of a variety of mammals, identifying some common characteristics.
	3: What do we know about the life cycles of amphibians?	Explain the differences in the life cycles of a mammal, an amphibian, an insect and a bird	Reporting and presenting findings from enquiries, including conclusions, causal relationships and explanations of and degree of trust in results, in oral and written forms such as displays and other presentations	Using a wide range of secondary sources of information	In this lesson children deepen their knowledge about the group of animals called amphibians. They find out about the life cycles of a variety of amphibians, identifying some common characteristics including the process of metamorphosis.

Module number and name	Lesson number and name	National curriculum links	Working scientifically links	Scientific enquiry type	Lesson summary
	4: What do we know about the life cycles of insects?	Explain the differences in the life cycles of a mammal, an amphibian, an insect and a bird	Reporting and presenting findings from enquiries, including conclusions, causal relationships and explanations of and degree of trust in results, in oral and written forms such as displays and other presentations	Using a wide range of secondary sources of information	In this lesson children deepen their knowledge about the group of animals called insects. They find out about the life cycles of a variety of insects, identifying some common characteristics.
	5: What do we know about the life cycles of birds?	Explain the differences in the life cycles of a mammal, an amphibian, an insect and a bird	Reporting and presenting findings from enquiries, including conclusions, causal relationships and explanations of and degree of trust in results, in oral and written forms such as displays and other presentations	Using a wide range of secondary sources of information	In this lesson children deepen their knowledge about the group of animals called birds. They find out about the life cycles of a variety of birds, identifying some common characteristics.
	6: What makes a successful life cycle?	Explain the differences in the life cycles of a mammal, an amphibian, an insect and a bird	Identifying scientific evidence that has been used to support or refute ideas or arguments	Finding things out using secondary sources of information	In this lesson children apply their knowledge and understanding of animal life cycles to an unfamiliar context. They invent their own animal, describe in detail each stage of its life cycle and explain how this will ensure its long-term success.
	7: How are humans helping endangered animals to complete their life cycles?	Explain the differences in the life cycles of a mammal, an amphibian, an insect and a bird	Identifying scientific evidence that has been used to support or refute ideas or arguments	Finding things out using secondary sources of information	In this lesson children find out about the ways in which humans are using science to help endangered animals complete their life cycles.
	EL1: Why do animals make incredible journeys as part of their life cycles?	Explain the differences in the life cycles of a mammal, an amphibian, an insect and a bird	Identifying scientific evidence that has been used to support or refute ideas or arguments	Finding things out using a wide range of secondary sources of information	In this lesson children find out about the incredible journeys that are undertaken by different types of animals during their life cycles.
Year 5 Module 2 Reproduction in Plants and Animals	1: How do flowering plants reproduce?	Describe the life process of reproduction in some plants and animals	Reporting and presenting findings from enquiries, including conclusions, causal relationships and explanations of and degree of trust in results, in oral and written forms such as displays and other presentations	Grouping and classifying	In this lesson children revise work about the part that flowers play in the life cycle of flowering plants. They learn about the role of the flower, its parts and their function, and the processes of pollination and fertilisation.
	2: Are all flowers on all plants the same?	Describe the life process of reproduction in some plants	Identifying scientific evidence that has been used to support or refute ideas or arguments	Grouping and classifying	In this lesson children further develop their understanding of the role of flowers in the reproductive cycle of plants.
	3: Do all plants reproduce by producing seeds?	Describe the life process of reproduction in some plants and animals	Reporting and presenting findings from enquiries, including conclusions, causal relationships and explanations of and degree of trust in results, in oral and written forms such as displays and other presentations	Finding things out using a wide range of secondary sources of information	In this lesson children learn about asexual reproduction, that is, the ways that plants can produce new plants from different parts of the parent plant, rather than by producing seeds.

Module number and name	Lesson number and name	National curriculum links	Working scientifically links	Scientific enquiry type	Lesson summary
	4: How do amphibians and insects reproduce?	Describe the life process of reproduction in some plants and animals	Reporting and presenting findings from enquiries, including conclusions, causal relationships and explanations of and degree of trust in results, in oral and written forms such as displays and other presentations	Finding things out using a wide range of secondary sources of information	In this lesson children find out in more detail about how amphibians and insects reproduce. They compare the process of reproduction in amphibians and insects, identifying and describing similarities and differences between the two and recognising both as examples of sexual reproduction, with some exceptions.
	5: How do mammals and birds reproduce?	Describe the life process of reproduction in some plants and animals	Reporting and presenting findings from enquiries, including conclusions, causal relationships and explanations of and degree of trust in results, in oral and written forms such as displays and other presentations	Grouping and classifying	In this lesson children find out more about how mammals and birds reproduce. The compare the process of reproduction in mammals and birds, identifying and describing similarities and differences between the two and naming both as examples of sexual reproduction.
	6: How does the human life cycle compare with that of other mammals?	Describe the changes as humans develop to old age	Recording data and results of increasing complexity using scientific diagrams and labels, classification keys, tables, scatter graphs and bar and line graphs	Noticing patterns	In this lesson children identify the stages of the human life cycle, including puberty and pregnancy, and compare lengths of gestation for different mammals.
	7: How do girls become women?	Describe the changes as humans develop to old age	Reporting and presenting findings from enquiries, including conclusions, causal relationships and explanations of and degree of trust in results, in oral and written forms such as displays and other presentations	Grouping and classifying	In this lesson children learn about the life cycle stage of puberty in girls.
	8: How do boys become men?	Describe the changes as humans develop to old age	Reporting and presenting findings from enquiries, including conclusions, causal relationships and explanations of and degree of trust in results, in oral and written forms such as displays and other presentations	Grouping and classifying	In this lesson children learn about the life cycle stage of puberty in boys.
Year 5 Module 3 Get Sorted	1: How can we compare and group materials?	Compare and group together everyday materials based on evidence from comparative and fair tests, including hardness, solubility, transparency, conductivity (electrical and thermal) and response to magnets	Recording data and results of increasing complexity using scientific diagrams and labels, classification keys, tables, scatter graphs, and bar and line graphs	Grouping and classifying	In this lesson children identify, compare and group materials based on their properties and according to their own or given criteria.
	2: Is a solid always hard?	Compare and group together everyday materials based on evidence from comparative and fair tests, including hardness, solubility, transparency, conductivity (electrical and thermal) and response to magnets	Reporting and presenting findings from enquiries, including conclusions, causal relationships and explanations of and degree of trust in results in oral and written forms such as displays and other presentations	Carrying out comparative and fair tests	In this lesson children investigate solids and compare them according to their properties.

Module number and name	Lesson number and name	National curriculum links	Working scientifically links	Scientific enquiry type	Lesson summary
	3: Is a liquid always runny?	Compare and group together everyday materials based on evidence from comparative and fair tests, including hardness, solubility, transparency, conductivity (electrical and thermal) and response to magnets	Planning different types of scientific enquiries to answer questions, including recognising and controlling variables where necessary	Grouping and classifying	In this lesson children carry out various comparative tests, exploring the viscosity of liquids.
	4: Are all metals the same?	Compare and group together everyday materials based on evidence from comparative and fair tests, including hardness, solubility, transparency, conductivity (electrical and thermal) and response to magnets	Identifying scientific evidence that has been used to support or refute ideas	Grouping and classifying	In this lesson children explore the ways in which metals are used around their school and in the wider world, and link these uses to the properties of the metals.
	5: Are all plastics the same?	Compare and group together everyday materials based on evidence from comparative and fair tests, including hardness, solubility, transparency, conductivity (electrical and thermal) and response to magnets	Planning different types of scientific enquiries to answer questions, including recognising and controlling variables where necessary	Grouping and classifying	In this lesson children identify and investigate the wide-ranging properties of plastics.
	6: To bounce or not to bounce: Why are sports balls so different?	Compare and group together everyday materials based on evidence from comparative and fair tests, including hardness, solubility, transparency, conductivity (electrical and thermal) and response to magnets	Planning different types of scientific enquiries to answer questions, including recognising and controlling variables where necessary	Carrying out comparative and fair tests	In this lesson children investigate the variables that affect how a ball bounces.
Year 5 Module 4 Everyday Materials	1: Which materials are used in our school buildings, what for and why?	Give reasons, based on evidence from comparative and fair tests, for specific uses of everyday materials, including metals, wood and plastic	Reporting and presenting findings from enquiries, including conclusions, causal relationships and explanations of and degree of trust in results, in oral and written forms such as displays and other presentations	Grouping and classifying	In this lesson children identify a variety of materials in different forms, observing how they are used for specific purposes within school buildings.
	2: Weighty problem: Which is the best carrier bag?	Give reasons, based on evidence from comparative and fair tests, for specific uses of everyday materials, including metals, wood and plastic	Planning different types of science enquiries to answer questions, including recognising and controlling variables where necessary	Carrying out comparative and fair tests	In this lesson children plan and carry out a fair test investigation into different types of plastic carrier bags, building on a lesson where they sorted, grouped and tested a wide range of plastics according to their properties.
	3: Which is the best type of plate to use?	Give reasons, based on evidence from comparative and fair tests, for specific uses of everyday materials, including metals, wood and plastic	Planning different types of science enquiries to answer questions, including recognising and controlling variables where necessary	Carrying out comparative and fair tests	In this lesson children carry out a comparative test to investigate how the properties of materials that are used to make plates affect their suitability for use in different situations or contexts.

Module number and name	Lesson number and name	National curriculum links	Working scientifically links	Scientific enquiry type	Lesson summary
	4: Cool box conundrum: Can the same container keep cold things cold and hot things hot?	Give reasons, based on evidence from comparative and fair tests, for specific uses of everyday materials, including metals, wood and plastic	Taking measurements, using a wide range of scientific equipment, with increasing accuracy and precision, and taking repeat readings when appropriate	Carrying out comparative and fair tests	In this lesson children investigate how a cool bag affects the temperature of hot and cold food.
	5: Mystery material: What will happen if we add water to the material?	Give reasons, based on evidence from comparative and fair tests, for specific uses of everyday materials, including metals, wood and plastic	Taking measurements, using a wide range of scientific equipment, with increasing accuracy and precision, and taking repeat readings when appropriate	Observing changes over different periods of time	In this lesson children observe and measure the effects of adding increasing volumes of water to quantities of a mystery material.
	6: Nappy ending: What's the best brand of nappy?	Give reasons, based on evidence from comparative and fair tests, for specific uses of everyday materials, including metals, wood and plastic	Identifying evidence that has been used to support or refute ideas or arguments	Carrying out comparative and fair tests	In this lesson children investigate different brands of nappies, coming up with their own questions and methods of enquiry. They identify the evidence that they need to collect so that they can provide information to parents about the various brands of nappy on offer and the brand claims.
	EL1: Are all bikes the same?	Give reasons, based on evidence from comparative and fair tests, for specific uses of everyday materials, including metals, wood and plastic	Reporting and presenting findings from enquiries, including conclusions, causal relationships and explanations of and degree of trust in results, in oral and written forms such as displays and other presentations	Grouping and classifying	In this lesson children identify the variety of materials (and their properties) that are used in making bicycles of different kinds.
	EL2: Spencer Silver and sticky notes: What's the stickiest glue?	Give reasons, based on evidence from comparative and fair tests, for specific uses of everyday materials, including metals, wood and plastic	Using test results to make predictions to set up further comparative and fair tests	Carrying out comparative and fair tests	In this lesson children learn about the chemist Spencer Silver and how he created Post-it™ notes almost by accident, as he worked to create a super-sticky glue.
Year 5 Module 5 Marvellous Mixtures	1: How can we separate mixtures?	Use knowledge of solids, liquids and gases to decide how mixtures might be separated, including through filtering, sieving and evaporating	Planning different types of scientific enquiries to answer questions, including recognising and controlling variables where necessary	Grouping and classifying	In this lesson children are introduced to the idea that materials can mix in different ways and that they can be separated. They make their own sieves to separate a complex mixture of dry solids.
	2: What happens when we mix liquids and solids?	Know that some materials will dissolve in liquid to form a solution, and describe how to recover a substance from a solution	Using test results to make predictions to set up further comparative and fair tests	Grouping and classifying	In this lesson children investigate dissolving solids.
	3: What makes a difference to how fast sugar or salt dissolves?	Know that some materials will dissolve in liquid to form a solution, and describe how to recover a substance from a solution	Planning different types of scientific enquiries to answer questions, including recognising and controlling variables where necessary	Planning comparative and fair tests	In this lesson children investigate what makes a difference to how rapidly a solid dissolves.
	4: How can we get drinkable water from seawater?	Use knowledge of solids, liquids and gases to decide how mixtures might be separated, including through filtering, sieving and evaporating	Planning different types of scientific enquiries to answer questions, including recognising and controlling variables where necessary	Observation over time	In this lesson children use their knowledge of evaporation and condensation to work out how to get materials back from a solution by investigating a real world problem: how to produce drinkable water from seawater, using limited equipment.

Module number and name	Lesson number and name	National curriculum links	Working scientifically links	Scientific enquiry type	Lesson summary
	5: How can we purify materials?	Use knowledge of solids, liquids and gases to decide how mixtures might be separated, including through filtering, sieving and evaporating	Planning different types of scientific enquiries to answer questions, including recognising and controlling variables where necessary	n/a	In this lesson children draw on the work they have done in Lessons 1–4 in order to consolidate their understanding of separating mixtures. They are challenged to develop their own methods to separate pure salt from a rock salt mixture.
	EL1: What will happen if we add a sprinkle of salt to a combination of liquids?	Understand that some materials will dissolve in liquid to form a solution, and describe how to recover a substance from a solution	Reporting and presenting findings from enquiries, including conclusions, causal relationships and explanations of and degree of trust in results, in oral and written forms such as displays and other presentations	Observation over time	In this lesson children explore what happens when oil and lemonade mix and the effect of a sprinkle of salt on the combination.
	EL2: How can we clean up contaminated water?	Use knowledge of solids, liquids and gases to decide how mixtures might be separated, including through filtering, sieving and evaporating	Reporting and presenting findings from enquiries, including conclusions, causal relationships and explanations of and degree of trust in results, in oral and written forms such as displays and other presentations	n/a	In this lesson children use their knowledge of separating mixtures to help them solve a real world problem.
Year 5 Module 6 Materials: All Change	1: Are the changes that happen around us reversible or non-reversible?	Demonstrate that dissolving, mixing and changes of state are reversible changes. Explain that some changes result in the formation of new materials and that this kind of change is not usually reversible, including changes associated with burning and the action of acid on bicarbonate of soda	Reporting and presenting findings from enquiries, including conclusions, causal relationships and explanations of and degree of trust in results, in oral and written forms such as displays and other presentations	Grouping and classifying	In this lesson children begin to explore how materials change when they are brought together in different ways. They identify types of changes and group them according to whether they think the change could be reversed, and then according to the conditions needed to bring about the change.
	2: How much gas can be produced by non-reversible change?	Explain that some changes result in the formation of new materials, and that this kind of change is not usually reversible, including changes associated with burning and the action of acid on bicarbonate of soda	Using test results to make predictions to set up further comparative and fair tests	Carrying out comparative and fair tests	In this lesson, as an example of a non-reversible change, children explore a variety of solids and liquids that react chemically when they are mixed.
	3: How long does it take for iron nails to rust?	Explain that some changes result in the formation of new materials and that this kind of change is not usually reversible, including changes associated with burning and the action of acid on bicarbonate of soda	Planning different types of scientific enquiry to answer questions, including recognising and controlling variables where necessary	Observing over time	In this lesson children set up an investigation to observe the changes that take place when some metals are exposed to the air or water.
	4: What happens when a candle burns?	Explain that some changes result in the formation of new materials and that this kind of change is not usually reversible, including changes associated with burning and the action of acid on bicarbonate of soda	Reporting and presenting findings from enquiries, including conclusions, causal relationships and explanations of and degree of trust in results, in oral and written forms such as displays and other presentations	Observing over time	In this lesson children observe and discuss the changes involved in burning a candle, recognising that there are reversible and non-reversible changes involved in the process.

Module number and name	Lesson number and name	National curriculum links	Working scientifically links	Scientific enquiry type	Lesson summary
	5: How long does it take for things to rust?	Explain that some changes result in the formation of new materials and that this kind of change is not usually reversible, including changes associated with burning and the action of acid on bicarbonate of soda	Reporting and presenting findings from enquiries, including conclusions, causal relationships and explanations of and degree of trust in results, in oral and written forms such as displays and other presentations	Observing changes over different periods of time	In this lesson children collate the results of the observation enquiries begun a couple of weeks before in Lesson 3, draw conclusions and present them to their peers.
	EL1: What would make the best rocket fuel?	Explain that some changes result in the formation of new materials, and that this kind of change is not usually reversible, including changes associated with burning and the action of acid on bicarbonate of soda	Planning different types of scientific enquiry to answer questions, including recognising and controlling variables where necessary	Comparative and fair tests	In this lesson children use knowledge gained from Lesson 2 to investigate a non-reversible change that takes place when an effervescent vitamin C tablet and water are combined.
	EL2: What are the bubbles in honeycomb toffee?	Explain that some changes result in the formation of new materials and that this kind of change is not usually reversible, including changes associated with burning and the action of acid on bicarbonate of soda	Reporting and presenting findings from enquiries, including conclusions, causal relationships and explanations of and degree of trust in results, in oral and written forms such as displays and other presentations	Observation over time	In this lesson children observe the process of making honeycomb toffee and identify the changes that happen to the materials used in the recipe.
Year 5 Module 7: Feel the Force	1: How can we measure forces?	Identify the effects of air resistance, water resistance and friction, which act between moving surfaces	Taking measurements, using a range of scientific equipment, with increasing accuracy and precision, including taking repeat readings when appropriate	Noticing patterns	In this lesson children extend their understanding of friction by learning how to measure forces using a Newton meter.
	2: Why does an object fall?	Explain that unsupported objects fall towards the Earth because of the force of gravity acting between the Earth and the falling object, and identify the effects of air resistance, water resistance and friction, which act between moving surfaces	Identifying scientific evidence that has been used to support or refute ideas or arguments	Carrying out comparative and fair tests	In this lesson children identify how scientific evidence is used to support and refute ideas, testing the explanations of Aristotle and Galileo about how things fall. Children investigate and find evidence for these ideas, exploring gravity as a non-contact force.
	3: What makes things move?	Explain that unsupported objects fall towards the Earth because of the force of gravity acting between the Earth and the falling object	Planning different types of scientific enquiries to answer questions, including recognising and controlling variables where necessary	Carrying out simple comparative and fair tests	In this lesson children investigate how forces make things change direction, speed up, slow down, start or stop moving and use force arrows to represent these.
	4: How can we slow down falling objects?	Explain that unsupported objects fall towards the Earth because of the force of gravity acting between the Earth and the falling object, and identify the effects of air resistance, water resistance and friction, which act between moving surfaces	Using test results to make predictions to set up further comparative and fair tests	Carrying out simple comparative and fair tests	In this lesson children plan and carry out a fair test investigation into air resistance, using parachutes. They make parachutes and measure the time taken for the parachute to fall. They will then use the results of their initial investigations to predict how they could improve their parachutes and plan and test out their ideas.
	5: Does the shape of an object affect its movement in a liquid?	Identify the effects of air resistance, water resistance and friction, which act between moving surfaces	Taking measurements, using a range of scientific equipment, with increasing accuracy and precision, taking repeat readings when appropriate	Carrying out comparative and fair tests	In this lesson the children learn that water resistance is a form of friction that opposes movement in water. They explore how the shape of an object affects its movement through a liquid.

Module number and name	Lesson number and name	National curriculum links	Working scientifically links	Scientific enquiry type	Lesson summary
	6: Do all heavy things sink?	Identify scientific evidence that has been used to support or refute ideas in arguments	Reporting and presenting findings from enquiries, including conclusions, causal relationships and explanations of and degree of trust in results, in oral and written forms such as displays and other presentations	Carrying out comparative and fair tests	In this lesson children measure the effect of upthrust on objects in water, by measuring and comparing weights of objects in water and air. They find out how the relationship between weight and size affects floating.
	7: How far can you stretch?	Explain that unsupported objects fall towards the Earth because of the force of gravity acting between the Earth and the falling object	Recording data and results of increasing complexity using scientific diagrams and labels, classification keys, tables, scatter graphs, and bar and line graphs	Noticing patterns	In this lesson children investigate what happens to rubber bands and springs when a force is applied.
	8: How can we use levers to help us?	Recognise that some mechanisms, including levers, pulleys and gears, allow a smaller force to have a greater effect	Taking measurements, using a range of scientific equipment with increasing accuracy and precision, including taking repeat readings when appropriate	Carrying out comparative simple and fair tests	This lesson introduces mechanisms – devices that change the effect of a force. Children investigate levers for moving things and increasing/decreasing a force.
	9: How can we lift a heavy load?	Recognise that some mechanisms, including levers, pulleys and gears, allow a smaller force to have a greater effect	Reporting and presenting findings from enquiries, including conclusions, causal relationships and explanations of and degree of trust in results, in oral and written forms such as displays and other presentations	Noticing patterns	In this lesson children use pulleys to lift objects.
	10: Can a wheel with teeth make work easier?	Recognise that some mechanisms, including levers, pulleys and gears, allow a smaller force to have a greater effect	Recording data and results of increasing complexity using scientific diagrams and labels, classification keys, tables, scatter graphs, and bar and line graphs	Noticing patterns	In this lesson children learn about gears, a third type of mechanism that helps us to do things by changing the effect of forces. Children identify where gears are used in everyday life.
Year 5 Module 8: The Earth and Beyond	1: What's in space?	Describe the movement of the Earth and other planets in the solar system relative to the Sun	Recording data and results of increasing complexity using scientific diagrams and labels, classification keys, tables, scatter graphs, bar and line graphs	Finding things out using a wide range of secondary sources of information	In this lesson children make observations of the night sky. Using secondary sources of information they consider explanations for, and raise questions about, their observations. They find answers to some of their questions through a 'journey into space', during which they explore diagrams and photographs of the solar system and beyond.
	2: What is a year?	Describe the movement of the Earth and other planets in the solar system relative to the Sun	Reporting and presenting findings from enquiries, including conclusions, causal relationships and explanations of and degree of trust in results, in oral and written forms such as displays and other presentations	Finding things out using a wide range of secondary sources of information	In this lesson children will draw a large 'plan' of the solar system and an annotated scientific diagram of Earth's orbit which they use to explain the year, number of days in a year, leap years and how astronomers in the past used the stars as markers for the start and finish of an orbit.
	3: What is a day?	Use the Earth's rotation to explain day and night and the apparent movement of the Sun across the sky	Reporting and presenting findings from enquiries, including conclusions, causal relationships and explanations of and degree of trust in results, in oral and written forms such as displays and other presentations	Noticing patterns	In this lesson children investigate how the Earth's rotation causes the apparent movement of the Sun across the sky.
	4: How does the Sun help us to measure time?	Use the idea of the Earth's rotation to explain day and night and the apparent movement of the Sun across the sky	Taking measurements, using a range of scientific equipment, with increasing accuracy and precision, taking repeat readings when appropriate	Observing changes over different periods of time	In this lesson children test different types of shadow clock. Children record the position and length of a shadow.

Module number and name	Lesson number and name	National curriculum links	Working scientifically links	Scientific enquiry type	Lesson summary
	5: What time is it around the world?	Use the idea of the Earth's rotation to explain day and night and the apparent movement of the Sun across the sky	Recording data and results of increasing complexity using scientific diagrams and labels, classification keys, tables, scatter graphs, bar and line graphs	Finding things out using a wide range of secondary sources of information	In this lesson children use a globe and world maps to find out about world time zones and how time is linked to longitude.
	6: Why do we have seasons?	Describe the movement of the Earth, and other planets, relative to the Sun in the solar system	Recording data and results of increasing complexity using scientific diagrams and labels, classification keys, tables, scatter graphs, and bar and line graphs	Observing change over time (modelled)	In this lesson children explore how Earth's tilt on its axis causes seasonal changes and changes in daylight hours.
	7: What are our conclusions about sunrise and sunset times?	Use the idea of the Earth's rotation to explain day and night and the apparent movement of the Sun across the sky	Identifying scientific evidence that has been used to support or refute ideas or arguments	Finding things out using a wide range of secondary sources of information	This lesson develops children's learning on time and seasons through investigating and explaining changes in the times of sunrise and sunset in different parts of the UK and different parts of the world.
	8: Why does the Moon change shape?	Describe the movement of the Moon relative to the Earth	Using test results to make predictions to set up further comparative and fair tests	Observing changes over different periods of time	In this lesson the children use their Moon diaries as as a source of information to investigate how the Moon appears to change shape over a month.

RESOURCE MATRIX

Year 5 Our Changing World	Resources
1: What signs of plant reproduction can we observe around our school?	Digital camera or iPad, magnifiers, sources of plant identification (e.g. FSC resources or similar)
2: How can we grow more plants without using seeds?	Variety of plant materials such as bulbs (autumn), rhizomes or tubers (spring), or stem cuttings from existing plants (summer) depending on the time of year, plants, e.g. fuchsias, geraniums, begonias, herbs such as rosemary and mint, bulbs for spring flowering, potato tubers, strawberry plants, lily, iris and gladioli rhizomes
3: Which plants are best to plant in our growing space? How can we ensure that produce is ready at the right time?	A space with good soil for growing, or large containers and grow bags as an alternative, garden tools, compost or manure, seeds (and their packets) for salad crops (e.g. lettuce, spring onion, rocket, nasturtium, pot marigold, cherry tomato, cucumbers, peppers) and flowering plants, potato tubers, strawberry plants, courgette or marrow plants
4: How can we ensure that plants in our growing space yield as many crops as possible?	A space with good soil for growing, or large containers and grow bags as an alternative, garden tools, compost or manure, seeds (and their packets) for salad crops (e.g. lettuce, spring onion, rocket, nasturtium, pot marigold, cherry tomato, cucumbers, peppers) and flowering plants, potato tubers, strawberry plants, courgette or marrow plants

Year 5 Module 1 Circle of Life	Resources
1: What is a life cycle?	Mini whiteboards, sticky notes, secondary sources for research, including quality non-fiction books, web-based resources, educational CDs, smartphone and tablet Apps, identification guides and leaflets
2: What do we know about the life cycles of mammals?	Mini whiteboards, secondary sources for research, including quality non-fiction books, web-based resources, educational CDs, smartphone and tablet Apps, identification guides and leaflets
3: What do we know about the life cycles of amphibians?	Mini whiteboards, secondary sources for research, including quality non-fiction books, web-based resources, educational CDs, smartphone and tablet Apps, identification guides and leaflets
4: What do we know about the life cycles of insects?	Secondary sources for research, including quality non-fiction books, web-based resources, educational CDs, smartphone and tablet Apps, identification guides and leaflets
5: What do we know about the life cycles of birds?	Secondary sources for research, including quality non-fiction books, web-based resources, educational CDs, smartphone and tablet Apps, identification guides and leaflets
6: What makes a successful life cycle?	Secondary sources for research, including quality non-fiction books, web-based resources, educational CDs, smartphone and tablet Apps, identification guides and leaflets
7: How are humans helping endangered animals to complete their life cycles?	Secondary sources for research, including quality non-fiction books, web-based resources, CDs, smartphone and tablet Apps, identification guides and leaflets, www.konicaminolta.com/kids/endangered_animals
EL1: Why do animals make incredible journeys as part of their life cycles?	Secondary sources for research, including quality non-fiction books, web-based resources, educational CDs, smartphone and tablet Apps, identification guides and leaflets

Year 5 Module 2 Reproduction in Plants and Animals	Resources
1: How do flowering plants reproduce?	Enough flowers for at least one between two children (ensure that the flowers are large enough to have identifiable male and female organs, such as alstroemeria or daffodils) magnifiers, digital microscopes, iPads, digital cameras
2: Are all flowers on all plants the same?	A variety of flowers (different from those observed in Lesson 1, including some single sex flowers), such as courgette, marrow, holly. If none are available, use images of single sex flowers, magnifiers, digital microscopes, digital cameras, modelling clay, clay, junk modelling resources
3: Do all plants reproduce by producing seeds	Examples of bulbs such as garlic, onions or shallots (some of which can be cut up),tubers, rhizomes, seed potatoes, plants in pots such as fuchsia, begonia, geranium, rosemary, mint, strawberry
4: How do amphibians and insects reproduce?	
5: How do mammals and birds reproduce?	
6: How does the human life cycle compare with that of other mammals?	Mini whiteboards and pens
7: How do girls become women?	Large sheets of paper and coloured pens or pencils for poster making; video camera, tablet computer with camera, or sound recording equipment, if available
8: How do boys become men?	Large sheets of paper and pens

Year 5 Module 3 Get Sorted	Resources
1: How can we compare and group materials?	Sticky notes, large sheets of paper, familiar classroom objects, for example, marker pen, pencil, paper clip, plant pot, sweatshirt, sports shoe, stapler, ruler, water bottle, lunch box, eraser; real objects and substances, for example, milk, shaving foam, ketchup, butter, yoghurt, jelly, hair gel, steam, sand, flour, sugar
2: Is a solid always hard?	Microscope, marshmallows and jelly sweets, chocolate buttons, cheese strings, cooked pasta, foil, elastic, net (or old tights), sponge, polystyrene, sand, soil, butter, brick, wooden ruler, plastic toy, metal object, piece of fabric, glass bottle, sponge, corn flour, water, tray or large bowl
3: Is a liquid always runny?	Large sheets of paper, honey, cooking oil, syrup, milk, washing up liquid, bubble bath, lemonade, yoghurt, different brands of tomato ketchup, wipe-clean ramps, whiteboards, teaspoons, tablespoons, stop watches or watches with second hands
4: Are all metals the same?	Magnets, examples of objects made of metals, for example, cooking pan, spoon, bell, paper clips, stepladder, power cable, access to books or the internet for research
5: Are all plastics the same?	Large bowl or jug, variety of large serving spoons made out of plastic, wood or metal, collection of objects made of plastics, for example, plastic bottles and packaging, plastic jugs and bowls (polythene), clothing made of polyester, strong ropes, washing line (nylon), beakers, plates, disposable cutlery, yoghurt pots (hardened polystyrene), insulation and packaging materials (expanded polystyrene), perspex sheets, lenses in torches (acrylic), pencils of different hardness, polystyrene cup, lemonade bottle, shampoo bottle, carrier bags, cling film, dustbin, washing up bowl or classroom tray, access to the internet or books for further research
6: To bounce or not to bounce? Why are sports balls so different?	Collection of balls, for example, cricket, tennis, hockey, snooker, football, rugby, basketball, volleyball, sponge, ping pong, golf, bowling, large hoops

Year 5 Module 4 Everyday Materials	Resources
1: Which materials are used in our school buildings, what for and why?	
2: Weighty problem: Which is the best carrier bag?	Lengths of thick dowel, broom handles, etc., modelling clay, large masses, for example, bricks, heavy books or cans of food to test bags, stop watches, different types of carrier bags, thick and thin plastic
3: Which is the best type of plate to use?	A variety of plates made of different materials (as similar in size as possible), ceramic, glass, pyrex, metal, plastic (different types and thicknesses), paper/card (different types and thicknesses), wood, plates that children can test to destruction (no best china), tools for chip test, safety goggles, tomato ketchup or similar for stain test, electronic weighing scales
4: Cool box conundrum: Can the same container keep cold things cold and hot things hot?	Thermometers, data loggers with temperature probes, hot water or soup in plastic containers with lids that have holes to allow access of thermometer or probe, ice cubes or ice cream in similar sized boxes or containers, cooked hot jacket potatoes, cool bags to use for testing, plus a couple of cool bags for disassembling
5: Mystery material: What will happen if we add water to the material?	Tub of 'Insta-Snow®' (available from TTS and other suppliers), water jugs, measuring cylinders, pipettes or water droppers, syringes, paper clips, jelly strings, hand lenses
6: Nappy ending: What's the best brand of nappy?	Mini whiteboards, water jugs, measuring cylinders, pipettes or water droppers, syringes, a collection of nappies with a variety of brands
EL1: Are all bikes the same?	A number of bikes of different types, mini whiteboards, bike catalogues, bike advertisements
EL2: Spencer Silver and sticky notes: what's the stickiest glue?	Glue spreaders, different kinds of glue that are safe for children to use, for example, stick glue, PVA glue; paste, fabrics and other materials, for example, plastic, cellophane, card, felt, hessian, lolly sticks, sand paper, sticky notes and Post-it™ notes, 'surfaces', for example, sheets of plastic, carpet tiles, lino, ceramic tiles, felt or fabric squares; 'labels', pre-cut rectangles of plastic or fabric, milk, gelatine, flour, access to the internet or books for further research

Year 5 Module 5 Marvellous Mixtures	Resources
1: How can we separate mixtures?	Disposable plates of different kinds – these can be pierced with nails, hole punches or bodkins to form makeshift sieves; selection of fabrics, nets and gauzes; Cupboard Catastrophe mixture – rice, raisins, large pasta, flour, dried lentils, dried peas, fine sand, white sugar, paperclips, wood shavings, plus three or four plastic spiders; large foil trays, plastic beakers, magnets, spoons
2: What happens when we mix liquids and solids?	Sand, salt, fruit syrup, brown sugar, large transparent beakers, collection of solids – powder paint, flour, sugar, sand, coffee granules, bath salts, tea leaves, baby powder, sugar substitute, bicarbonate of soda; collection of solvents – oil, vinegar, water; beakers, spoons, weighing equipment, measuring jugs
3: What makes a difference to how fast sugar or salt dissolves?	Rock salt, table salt, icing sugar, Demerara sugar, granulated sugar, water, disposable transparent beakers, saucers, teaspoons, measuring equipment, timers, hand lenses, mini microscopes
4: How can we get drinkable water from seawater?	Large bowls, saucers, salt solution, water jugs, desk lamps or other strong light sources, cling film, plastic sheeting
5: How can we purify materials?	Chunky rock salt with impurities, sand, gravel chips and soil; materials to create filter beds, such as felt, wood shavings, sand, insulation fibre, wadding, cotton wool, three 1-litre plastic lemonade bottles pre-cut at neck (these will be used to create filters that can be prepared by children in advance of the lesson – see Resource sheet 1), water, water jugs, selection of sieves and funnels of different sizes
EL1: What will happen if we add a sprinkle of salt to a combination of liquids?	Transparent pint beakers, cheapest vegetable oil, cheapest lemonade (in large quantities), food colour and water droppers, table salt, trays full of sand to stabilise beakers of liquid
EL2: How can we clean up contaminated water?	Containers of 'contaminated' water – a soupy mixture containing as nasty a mix as you like: oil, soil, sand, pebbles, leaves and twigs, bits of plastic, and so on, and some pond water for children completing Challenge 3. Sieves with different grade mesh, funnels of different size, filter papers, buckets, bowls, plastic sheeting, mop and bucket. Material collection for making filter beds – fine sand, gravel, wadding, felts and other thick fibrous fabrics , foil food trays, microscopes

Year 5 Module 6 Materials: All Change	Resources
1: Are the changes that happen around us reversible or non-reversible?	Small bottles of lemonade, shaving foam canisters, salt, water, chocolate buttons, beakers and small plates, paper towels
2: How much gas can be produced by non-reversible change?	Disposable latex gloves; solids: bicarbonate of soda, tartaric acid, baking powder, effervescent vitamin C tablets, effervescent indigestion tablets; liquids: water, white vinegar, lemon juice; small beakers, disposable cups, plastic teaspoons, milk bottles or cartons, small pop bottles
3: How long does it take for iron nails to rust?	Iron nails, metal paint, paint brushes, petroleum jelly (or similar thick grease), oil, salt, lemon juice, vinegar, lemonade, water, plastic disposable beakers (transparent), clingfilm, objects made of metal, including washers, key, spoons, copper wire, aluminium foil, tin can, zinc and copper nails
4: What happens when a candle burns?	Candles (see below), metal containers filled with sand, glass jars of varying size, paper, pencils, digital camera, mini whiteboards
5: How long does it take for things to rust?	The beakers containing iron and other materials that were being observed in different conditions to investigate rusting; the children's observation records
EL1: What would make the best rocket fuel?	Narrow measuring cylinders or small beakers, water, effervescent vitamin C tablets (one per group), small containers with snap-on lids. Stomp rocket (optional), sticky tack, water in jugs
EL2: What are the bubbles in honeycomb toffee?	For the honeycomb toffee: 100 g sugar, 2 tablespoons of syrup, 1 heaped teaspoon of bicarbonate of soda, 1 tablespoon of vegetable oil for oiling the pan, a large heavybased saucepan, a wooden spoon, a tin lined with aluminium foil (to save on washing up!) and a means of heating, such as a portable stove or access to the school kitchen. Challenge 3 – honeycomb toffee that has cooled, samples of pumice stone, sponge cake, bread, expanded polystyrene, hand lenses, access to sources of information about these materials

Year 5 Module 7 Feel the Force	Resources
1: How can we measure forces?	Newton meters (2.5 N, 5 N, 10 N, and 20 N), modelling clay, toy vehicles, three sizes of match box, some of which have different materials glued to the base (for example, rough sandpaper, aluminium foil, rectangular sections cut from rubber gloves or thin foam rubber, cotton cloth)
2: Why does an object fall?	Objects to drop to demonstrate something falling, empty camera film canisters, modelling clay, good quality cupcake cases, timers
3: What makes things move?	Bubble mixture, big toy vehicle, balloon, straws, fans (battery and paper), hair dryer, table tennis ball, card arrows of different sizes
4: How can we slow down falling objects?	Different types of string, scissors, plastic bin liners, different sized small plastic, a range of materials, including tissue paper, plastic, fabric, card, paper
5: Does the shape of an object affect its movement in a liquid?	Modelling clay, viscous children's bubble bath, 1000 ml measuring cylinders (or 1.5 litre plastic bottles with the necks cut off), timers (stopwatches or stopclocks), 1000 ml jugs, digital scales, kitchen roll, sticky notes, elastic bands, masking tape (if required for the plastic bottles)
6: Do all heavy things sink?	Fish tanks of water, Newton meters, modelling clay, balloon, large clear plastic tumblers, fruit and vegetables, weighing scales, salt, elastic bands
7: How far can you stretch?	Fixed cup hooks, selection of springs, paper clips, rubber bands, sets of hanging slotted weights up to 100 g, hanging weights of 50 g and 100 g, Newton meters, tape measures, modelling clay, small objects from around the classroom to hang on rubber bands
8: How can we use levers to help us?	Everyday objects that use class one levers (e.g. claw hammer, scissors, pliers, metal spoon), empty tins with inset metal lids, long-handled wooden spoons, 1 litre plastic bottles filled with water to weigh 1 kg, stiff cardboard tubes approximately 3 cm diameter, modelling clay, push/pull meters up to 10 N, books (at least thick paperback size), wooden ramps to be used in Challenge 1
9: How can we lift a heavy load?	Wooden dowel (at least 2 cm diameter) or brush handles, pulley sets or metal coat hangers and curtain rings to slide on dowel, cotton reels, string or thin rope, small bucket , sand, Newton meters, sets of slotted weights in 10 g denominations, hanging masses of 50 g, 200 g, 500 g and 1000 g.

Year 5 Module 7 Feel the Force	Resources
10: Can a wheel with teeth make work easier?	Balloon whisk, rotary whisk, 2 bowls and 2 egg whites, cheap clock with removable back, commercially produced plastic gear wheels, plastic bricks, axles, plastic brick bases

Year 5 Module 8 The Earth and Beyond	Resources
1: What's in space?	A2 paper
2: What is a year?	A big ball of string (at least 20 metres in length), a big block of chalk or a couple of packs of chalk (different colours , if possible), eight large balls, card strips about a third A4, each with the name of a planet
3: What is a day?	Globe, sticky tack, bright torch, cocktail stick, compass
4: How does the sun help us to measure time?	Large paper or polystyrene plates, permanent marker pens (different colours), fairly small but sharpened pencils, modelling clay, cereal boxes, scissors, direction compasses, watches, torches, small model figures (about 8–15 cm high)
5: What time is it around the world?	Globes, torches, sticky tack, large map of the world that shows major cities, online maps of the world that show longitude, internet world clock
6: Why do we have seasons?	Battery powered lanterns that shine in all directions (or torches), a globe, sticky tack, materials for making a poster, piece of dowelling, small ball of modelling clay, secondary sources for research
7: What are our conclusions about sunrise and sunset times?	Globe, torch, maps and atlases of the UK and the world, access to the internet for further research
8: Why does the Moon change shape?	A border strip of dark paper for a 'Moon phase' frieze, at least 30 circles for cutting out moon shapes, chalks, a big ball half covered with black plastic and half covered with a white plastic bag, a piece of black sugar paper per child, with a circle drawn in the middle of each sheet in white chalk, (number these sheets individually from 1, up to the number of children in the class), access to the internet to check online calendars, a calendar for the month ahead

TEACHING FRAMEWORK

OUR CHANGING WORLD

INTRODUCTION

In this module children develop their understanding of the life cycles of plants and of reproduction as a specific stage of those life cycles. This module links to Module 2, Reproduction in Plants and Animals. It provides children with further opportunities to explore ideas from that module actively and practically, applying knowledge, understanding and skills, often within the outdoor learning environment.

As they explore 'Our Changing World' of plants, children look for evidence of plant reproduction, for example, flowers, seed heads, berries and fruits on plants, throughout the year. They make observations of a wide variety of plants at different stages of their life cycles. They note all of this detail in an Our Changing World diary, on plant maps and also on planting plans. Children explore practically some of the methods of growing plants without seeds and propagation that they learned about in Lesson 3 of Module 2, Reproduction in Plants and Animals. They design and carry out a planting plan to grow a range of plants using seeds, bulbs, tubers, rhizomes, corms and leaf, stem and root cuttings, ready for a summer term 'produce sale' of crops. They investigate different ideas about how to improve crop yields and quality.

In working scientifically children carry out a range of enquiries, often making repeated observations in order to develop a deeper understanding of changes that take place over extended time periods (days, weeks and months) and through the seasons. Children notice, analyse and interpret patterns in data that they and others have collected. They use this evidence, coupled with that from secondary sources, to draw conclusions about what they have found out about the life cycles, and particularly the stage of reproduction, in plants in their locality and in plants that they cultivate.

Key vocabulary:

flower, carpel, stamen, pollen, seed, seed head, berry, hip, fruit, pollinator, pollination, fertilise, fertilisation, seed dispersal, male, female, organs, sex, propagate, propagation, stem/leaf/root cutting, runner, tuber, rhizome, bulb, crop, cropping, produce, yield, glut, names of fruit and vegetables being grown

FACT FILE:

Sexual reproduction in flowering plants

The reproductive organ of flowering plants is the flower.

The broad term 'flower' can be used to describe both **simple** and **compound** flowers. A simple flower has petals and contains a single set of reproductive parts at the centre, such as a buttercup or lily. Compound flowers appear to be single flowers, but the flower itself is actually made up of numerous small flowers arranged within a flower head. Daisies, dandelions and sunflowers are good examples of this. Most flowering plants have flowers with both male and female parts – 'perfect flowers' such as apple, tulip, daisy, dandelion and rose.

Some plants have separate male flowers and female flowers on the same plant, such as corn, courgette, marrow, squash and cucumber.

A smaller number of plants have male flowers and female flowers on separate plants, such as willow, maple and holly.

Children should learn that all plants do not produce 'perfect flowers' with both male and female organs, but that there are some plants with different sex flowers on the same or separate plants.

The female part of a flower consists of the **carpels**, which is where the seeds are formed. It has three parts: the **stigma**, the **style**, and the **ovary**. The male parts of the flower are the **stamens**, which produce **pollen**. Each stamen has two parts: an **anther** and a **filament**. The **anther** contains the pollen and the **filament** supports the anther.

When the flower is **pollinated**, a pollen grain sticks to the stigma. It then travels through a narrow tube which grows down through the style to the **ovary**. In the ovary, the pollen joins with the **ovules**. This fusion of the male and female cells is called **fertilisation** and the fused cells divide to develop into **seeds**. After fertilisation, the ovary usually swells and becomes the **fruit**.

Asexual reproduction in plants

Many plants can also reproduce without forming seeds. This is called asexual or vegetative reproduction, which results in new plants that are genetically identical to the parent.

Plants may reproduce themselves naturally:

• Below ground – rhizomes, tubers, bulbs and corms. These are underground growths on the root or stem of a plant and contain stores of food to provide for the growing young plant.

• Above ground – the parent plant produces runners and new plants sprout along its length.

Common misconceptions:

• Children may not recognise that reproduction is a characteristic of living things.

• Some children think that plants do not reproduce sexually at all.

• Children may think that bees and other insects visit flowers to pollinate them. They visit flowers to collect nectar; their role in pollination is accidental as far as the insect is concerned.

• Children may think that bees fertilise flowers; they pollinate them. Fertilisation happens when male and female genetic material fuses.

Big Cat book links

Star Gazing Celia Warren 978-0-00-746531-6 Band 12 Copper	A beautifully illustrated poetry book focuses on the delights of the natural world.
Jaws and Claws and things with Wings Valerie Bloom 978-0-00-746539-1 Band 14 Ruby	A poetry collection inspired by the strange and wonderful creatures in the natural world.

OUR CHANGING WORLD

 LESSON 1: WHAT SIGNS OF PLANT REPRODUCTION CAN WE OBSERVE AROUND OUR SCHOOL?

LESSON SUMMARY:

During these lessons children identify a variety of plants to observe, visit them regularly throughout the year and look for evidence of plant reproduction, for example, flowers, seed heads, berries and fruits on plants. They also record the numbers and types of pollinators they observe, for example, bees, butterflies and moths, at different times of the year. Video 1 can provide some background on how different plants disperse seeds.

By the end of these lessons children are able to describe the specific evidence of reproduction in plants that they have seen, identify patterns in how and when plants of the same or different species reproduce, and suggest reasons why this might be important.

Preparation required: Children need to draw (or be provided with) a simple outline map of the outdoor area around school to use in the Enquire part of the lesson. An outside visit should be organised if children are drawing this themselves. For the Reflect and review part of the lesson you need to collect a range of evidence of reproduction in plants, for example, a variety of flowers, seed heads, berries and fruits from plants, as you walk around the site with children.

These lessons build on learning from Lesson 1 of Module 2, Reproduction in Plants and Animals.

Key vocabulary:

reproduction, reproduce, flower, carpel, stamen, pollen, seed, seed head, berry, hip, fruit, pollinator, pollination, fertilisation, seed dispersal

Resources:

Digital camera or iPad, magnifiers, sources of plant identification (for example, FSC resources or similar)

Health and safety:

Always wash hands after handling plants, seeds and soil. Some plants may have sap that will irritate a child's skin. Some bulbs (for example, hyacinths) can also cause skin irritation.

Key information:

If the weather is sunny, warm and relatively still, children should see plenty of flying insects (for example, butterflies, bees and some beetles) visiting flowering plants and acting as pollinators. These insects, and other pollinators including birds, aid the pollination process by passing pollen from flower to flower, which ensures that fertilisation takes place.

National curriculum links:

Describe the life process of reproduction in some plants and animals

Learning intention:

To observe, record and collect evidence over time of life cycle changes to plants within the local environment

Scientific enquiry type:

Observing changes over different periods of time

Working scientifically links:

Recording data and results of increasing complexity using scientific diagrams and labels, classification keys, tables, scatter graphs, and bar and line graphs

Success criteria:

- I can make detailed observations of plants at different times of the year, noticing the stage that they have reached within their life cycle.
- I can look for patterns across plants of the same and different species (for example, are all the plants in flower or producing fruit?).
- I can suggest reasons for differences in how and when plants reproduce.

EXPLORE:

Show children the Reproduction in plants slideshow (Slideshow 1), which features a number of flowering plants and the different stages of their life cycles.

Ask: *What are we seeing evidence of here?*

Ensure that children use the term 'reproduction' in their descriptions of the different stages of a plant life cycle and other scientific vocabulary, for example, names of parts of the flower or processes such as pollination, fertilisation and seed dispersal.

Prompt them to expand on their ideas.

Ask: *What signs of reproduction in plants might you see as you make observations around the school? What else would indicate that reproduction is happening?*

Encourage children to think about the likely presence of pollinators, particularly flying insects.

ENQUIRE:

Tell children that they are going to take a walk around the school grounds (or a nearby green space). Provide them with a simple map of the area and ask them to take their Our Changing World diaries. They can also take digital cameras and iPads to record their observations.

Tell them that on the first visit they should identify a number of different flowering plants, bushes and trees: these same sites will be revisited regularly during the course of the year. Explain to the children that they should mark on their maps what they have seen and where, using plant identification sheets (for example, FSC resources or similar) to research the names and types of plants. Tell them to look out for insects and other pollinators, and make a note of the weather, how many pollinators they see and where they see them. Children record all their notes in their Our Changing World diaries.

Explain to children that on subsequent visits they will check for evidence of plant reproduction. For example, whether flowers have turned to fruit or whether seed heads have been produced and are dispersing seeds. Remind children to make notes about the pollinators that they see, to note which plants they saw them on and to record how numbers of pollinators vary, which may depend on the weather conditions and the time of year.

The challenges are differentiated by the level of detail children are required to observe and report, and how they classify plants according to different reproduction processes.

Challenge 1: Children identify different plants and pollinators

Support children to plan a route that includes a variety of different plants.

Ask: *What are the names of the plants? How could we find out? Do they have flowers at the moment? Do they produce fruits, berries or hips? Do they produce seed heads?*

Remind the children to look out for insect pollinators and draw their attention to plants on which pollinators are likely to be seen in higher numbers.

Challenge 2: Children make and record detailed observations of plants and pollinators

Tell the children to add as much detail to their maps as possible, planning a route that includes a variety of plants. Prompt them to note the names of plants, together with evidence of plant reproduction they have seen, and other observations. Remind the children to look out for insect pollinators and note which plants attract higher numbers of pollinators.

Challenge 3: Children identify specific types of plants according to how they are pollinated, and how they disperse seeds and reproduce

Tell the children that they should plan a route that includes as wide a variety of plants as possible. Identify some specific types of plants they need to locate, including: plants with flowers that are pollinated in different ways; plants that disperse their seed in different ways; plants that don't have flowers, but that reproduce in another way.

Remind them to record all their information in their Our Changing World diaries.

REFLECT AND REVIEW:

On returning to the classroom after each visit to the plants that children are monitoring, use a visualiser to project onto the whiteboard either images of evidence of plant reproduction from children's records or, preferably, actual parts of plants.

Ask children to talk with a partner, choose three objects from the selection and decide which one is the 'odd one out', explaining their reasoning to each other. They can then challenge other pairs to work out which they think is 'odd' and why. An example of this might be: a dandelion flower, a rose flower and a sycamore key. In this case the rose might be the odd one out because the seeds of the other two plants are dispersed by the wind. Alternatively, the sycamore key may be the odd one out because it is a seed head, whereas the other two are flowers.

EVIDENCE OF LEARNING:

Listen to children's responses as they select plants to observe regularly and then as they make repeat visits to look for evidence of the plants' life cycles over time. Look at their Our Changing World diaries: Do they select plants to visit systematically, ensuring that they include a variety of different types of plants? Can they identify evidence of the life cycle stage plants have reached, using appropriate scientific vocabulary to describe their observations? Do they notice patterns across plants of the same and different species (for example, are all plants of one kind in flower at the same time or producing fruit)? Do children note the different numbers (and types) of pollinators on different visits, identifying a link between, for example, the numbers of flying insects they see and the weather on that day or the numbers of flowers in full bloom? Can they use their observations to help identify differences in how and when plants of different kinds reproduce?

OUR CHANGING WORLD

LESSON 2: HOW CAN WE GROW MORE PLANTS, WITHOUT USING SEEDS?

LESSON SUMMARY:

During these lessons children explore practically some of the methods for growing new plants, which they learned about in Lesson 3 of Module 2, Reproduction in Plants and Animals.

By the end of these lessons children have successfully grown a variety of flowering plants and crops without using seeds, and have observed carefully the similarities and differences between propagating plants from seed and from other plant materials,for example, stem, leaf and root cuttings, runners, tubers, rhizomes and bulbs.

Preparation required: Video 1 is the video used in Module 2, Lesson 3 and is reused during the first lesson of this sequence. Children should revisit the sets of instructions that they produced during Lesson 3 of Module 2, Reproduction in Plants and Animals. These will help them as they begin to practically propagate plants for themselves.

Plants propagated during these lessons could be planted out by children in their growing place (see Lesson 3 and Lesson 4).

Key information:

Depending on the time of year, different methods of propagation can be attempted. For example, stem cuttings from wood of varying hardness can be taken at different times of the year, as can leaf cuttings. Root cuttings should be taken in the winter when the plants are dormant.

Key vocabulary:

reproduce, propagate, stem/leaf/root cutting, runner, tuber, rhizome, bulb

Resources:

Variety of plant materials such as bulbs (autumn), rhizomes or tubers (spring), or stem cuttings from existing plants (summer) depending on the time of year, plants, for example, fuchsias, geraniums and begonias, herbs such as rosemary and mint, bulbs for spring flowering, potato tubers, strawberry plants, lily, iris and gladioli rhizomes

Health and safety:

Wash hands after working with soil or potting compost or handling plant material. Some plants may have sap that will irritate a child's skin. Some bulbs (for example, hyacinths) can also cause skin irritation.

National curriculum links:

Describe the life process of reproduction in some plants and animals

Learning intention:

To observe first-hand how plants are able to reproduce themselves by using different parts of the parent plant to produce new plants

Scientific enquiry type:

Observing change over different periods of time

Working scientifically links:

Identifying scientific evidence that has been used to support or refute ideas or arguments

Success criteria:

- I can describe a variety of ways in which plants are able to reproduce, without using seeds.
- I can prepare plant material so that it has the best possible chance of growing into a new plant.
- I can make observations of plants as they grow, comparing different methods of growing new plants without seeds.

EXPLORE:

Show children Plant propagation (Video 1) which they first saw during Lesson 3 of Module 2, Reproduction in Plants and Animals.

Ask: *What different methods of plant reproduction and propagation did we hear about? What is the difference between natural plant reproduction processes and artificial propagation methods used by gardeners? Can you list the different types?*

Refer children back to the instructions that they produced during Lesson 3 of Module 2.

Ask: *Which methods should we use in these lessons to grow new plants without seeds? Which do you think would work the best? Why?*

Take feedback from children and ensure that as many of their ideas as possible are explored.

ENQUIRE:

Explain to children that during the course of these lessons they are going to have first-hand experience of a variety of methods for growing new plants without using seeds. The methods that they use will depend on the season.

The autumn term is a good time to plant spring-flowering bulbs and observe their development. This involves planting the bulbs, marking their positions and revisiting the site every week or so to check for signs of growth.

During the spring term children may plant rhizomes and tubers for summer flowering. For example, plant early potatoes in containers or bags during March/April. These should produce a crop of potatoes during the summer term.

Key information:

The ability for a plant to reproduce by means other than by producing seeds saves energy – although the process of seed production and dispersal ensures more genetic diversity within the population. In the natural environment plants may reproduce underground using rhizomes, tubers, bulbs and corms. These are underground growths on the root or stem of a plant that contain stores of food to provide for the growing young plant. Above ground the parent plant may produce runners, along the length of which new plants sprout. The new plants produced by these underground and above ground means of reproduction are exact genetic copies, or 'clones', of the parent.

Gardeners use techniques to reproduce plants from stem, root or leaf cuttings. When pieces of plant material are placed in soil or water new plants, complete with root systems, are produced. This is called propagation.

The summer term is a good time to identify runners on existing plants. Advise children that they should peg runners down to encourage new plants to root. Summer is also a time to propagate plants by taking cuttings, for example, from fuschia, geranium, mint and rosemary.

The challenges are differentiated by the extent to which children organise independently the growing of new plants without seeds, the detail with which they record their observations and the requirement for evaluating the success of the methods they use.

Challenge 1: Children keep a diary to record the growth of their new plants without seeds

Remind the children of the instructions for growing new plants without seeds that some of them produced during Lesson 3 of Module 2, Reproduction in Plants and Animals. Share these instructions with the class using a visualiser or photocopy them for children to use. Support the children, as necessary, as they prepare different kinds of plant material and get it ready to grow. They should make an initial sketch and record notes about the process in their Our Changing World diaries. Explain to the children that they then need to look for and record any signs of growth regularly. Prompt the children to look each time a change is visible, and ask them to draw and record what they can see in their Our Changing World Diaries.

Ask: *What can you see? What do you think is happening?*

Challenge 2: Children make and record systematic observations of their different methods of growing new plants without seeds

Explain to the children that they need to check for signs of growth periodically (how often, and what they observe will depend on the type of propagation being explored). For example, if stem cuttings are being rooted in water, children will be able to see the new roots starting to grow very quickly. It takes much longer to see signs of growth from cuttings in soil. However, one cutting might be carefully removed after 2 weeks, so that the root growth can be examined.

Ask: *What do you notice first? After how long? What can you see 2 weeks later? A month later? When do new leaves or flower buds appear?*

Encourage the children to draw what they see and make notes about what they notice, and how long it takes before changes are evident, in their Our Changing World diaries.

Challenge 3: Children use observation data collected over time to evaluate the effectiveness of different methods of growing new plants without seeds

Explain to the children that across this series of lessons they should compare the plants that they have grown without seed, referring to evidence in their Our Changing World diaries.

Ask: *Which method seems to have given the best results? What makes you say that?*

The children might explain that the plants seem stronger or that they have grown more quickly, or have begun to flower sooner and produce fruit/seeds more readily. Ask them to compare these methods for growing new plants with growing seeds.

Ask: *Why might a market gardener choose to take cuttings, rather than grow plants from seed?*

REFLECT AND REVIEW:

Challenge children to summarise what they have learned and to produce a Top Tips Guide for budding gardeners, on growing plants without seeds. Ask them to talk to their partners and to come up with two ideas that they think should feature in the Top Tips Guide. Take feedback from children and help them to combine their best ideas. The completed list could feature on the school website and be shared with children in other classes, as well as their parents.

EVIDENCE OF LEARNING:

Listen to children's responses as they develop the Top Tips Guide, and observe and listen to them as they approach the challenges (particularly those children tasked with Challenge 3). Can they describe different ways in which plants reproduce, other than by producing seed? Can they access or follow instructions for different methods for growing new plants without seeds, preparing plant material carefully so that it has the best possible chance of growing into a new plant? Do children make detailed observations of plants as they grow, making notes, for example, about the speed at which cuttings take root and produce a new plant? Can they compare the processes of growing new plants from seeds and using other methods of reproduction? Do they give useful advice to budding gardeners in their Top Tips Guide on growing plants without seeds?

OUR CHANGING WORLD

 LESSON 3: WHICH PLANTS ARE BEST TO PLANT IN OUR GROWING SPACE? HOW CAN WE ENSURE THAT PRODUCE IS READY AT THE RIGHT TIME?

LESSON SUMMARY:

During these lessons children apply their knowledge and understanding of plant life cycles as they plan for and grow plants across the year, ready for summer term 'produce sales' or similar events.

By the end of the learning sequence children have a greater understanding of the importance of knowing the length of a plant's life cycle, and have considered life cycle length as they made plans for produce to be generated, appreciating the need to avoid having a glut of one type of produce at any one time. Children also recognise that knowing the length of a plant's life cycle is especially important for commercial producers of flowers, vegetables and fruit.

National curriculum links:

Describe the life process of reproduction in some plants

Learning intention:

To investigate the length of a plant's life cycle, from planting to crop production, and use this information in planning and caring for our growing space

Scientific enquiry type:

Observing changes over different periods of time

Working scientifically links:

Planning different types of scientific enquiries to answer questions, including recognising and controlling variables where necessary

Success criteria:

- I can investigate the length of time different plants take to complete their life cycles and produce crops, recognising the importance of this to commercial crop growers.
- I can use information on the lengths of plant life cycles to help develop a plan for what plants to grow and when to plant them.
- I can work with others to plan, prepare plants and seeds, and plant out our growing space.

EXPLORE:

Explain to children that during these lessons they are going to plan what they might plant in a growing space to produce crops of fruit, vegetables or flowers.

Ask: *What do we need to think about if we are going to have our crops planted and grown, and their produce ready for sale by …? What sorts of crops should we grow? Vegetables, fruit, flowers? Can anybody suggest any plants that they think might give us 'produce' at the right time?*

Provide children with a wide range of seed packets to look at. (Copy some, if necessary, so that all the groups can see the information on each type of seed.) Explain to children that they are going to use the information to decide which plants would be most suitable to plant as crops. Prompt them to think about what they need to consider.

Ask: *When will the crops need to be planted? How long will they take to crop (for example, produce flowers, vegetables or fruit)? When will the produce be ready for sale? Which plants should provide the highest yield?*

Advise children that they should make notes about what they find out in their Our Changing World diaries.

ENQUIRE:

Explain to children that they need to use the information they gathered during the first part of the lesson to help produce a plan for their growing space. They need to think about space and timing: where to plant different varieties and when. Their aim is to plant and grow produce that is ready for harvesting throughout the summer term, either to eat or for sale.

Key information:

Children should make use of the new plants that they generated during Lesson 2, for example, potatoes grown from tubers, strawberry plants from runners, etc.

Key vocabulary:

crop, cropping, produce, yield, glut, names of fruit and vegetables being grown

Resources:

A space with good soil for growing (or large containers and grow bags as an alternative) garden tools, compost or manure, seeds (and their packets) for salad crops (for example, lettuce, spring onion, rocket, nasturtium, pot marigold, cherry tomato, cucumbers, peppers) and flowering plants, potato tubers, strawberry plants, courgette or marrow plants.

Health and safety:

Wash hands after working with soil or potting compost or handling plant material. Some plants may have sap that will irritate a child's skin. Some bulbs (for example, hyacinths) can also cause skin irritation.

After the crops have been planted, ensure that children water the seeds and plants regularly, removing weeds and looking for early signs of pests that might eat or damage the crops. (Lesson 4 includes a focus on pest control and crop care.)

The challenges are differentiated by the support given for planning, sequencing and identifying suitable plants.

Challenge 1: Children plan a planting sequence and layout to grow produce to sell or eat

Provide the children with the Crop planner grid (Resource sheet 1) that will help them to plan and organise a growing sequence, and a list of the plants (Resource sheet 2) that they might use in their growing space. Ask them to look at the information that they gathered in the first part of the lesson to help them to choose the plants and decide when they should be planted.

Challenge 2: Children choose suitable plants to grow to sell or eat, and plan a planning sequence and layout for their space

Provide the children with the Crop planner grid (Resource sheet 1) to help them to plan a sequence of planting and growing that will produce crops of different kinds through the summer term. If necessary, suggest examples of several crops that they might consider and ask the children to add others that they think would be good to grow and sell.

Challenge 3: Children choose suitable plants to grow and sell on a fortnightly basis throughout the summer term, and plan a planning sequence and layout

Challenge the children to plan for a sequence of crops across the summer term. Explain that their aim is to have something ready for sale once a fortnight from May to July. Remind the children to think about which crops are fast growing and which give a high yield, and to consider including some 'tempting' produce, such as cut flowers or strawberries.

Ask: *What problems could you face? How can you ensure the success of your growing space? What regular activities will you all need to undertake?*

REFLECT AND REVIEW:

Show children the Market garden video (Video 1), of an organic market garden where salad crops and vegetables are produced on a large scale. Ask children to listen out for the answers to a number of questions:

• How could we use what we have heard in the video to improve the success of our growing space?

• How could we improve the quality of our produce?

• Were there any clues about how pests and diseases can be managed, without the use of sprays or insecticides?

EVIDENCE OF LEARNING:

Listen to children as they consider which plants to grow in their growing space and to their responses during the Reflect and review part of the initial lesson in particular. Do they investigate successfully the length of time it takes different plants to produce crops, using and comparing information from seed packets and other information to do this? Are they able to use this information to help them to develop a plan for which plants to grow in the growing space? Do children recognise the importance of when to plant each crop, so that produce is not ready all at the same time? Do they work well with others to plan for planting, prepare plants and seeds, and plant out the growing space? Can they explain why the length of time a crop takes to grow is important to commercial crop growers?

OUR CHANGING WORLD

LESSON 4: HOW CAN WE ENSURE THAT PLANTS IN OUR GROWING SPACE YIELD AS MANY CROPS AS POSSIBLE?

LESSON SUMMARY:

During these lessons children set up and carry out an investigation to test a variable that may affect crop production. They apply their knowledge and understanding of plant life cycles, as they care for and grow plants across the year, ready for summer term 'produce sales', or similar.

By the end of the learning sequence children have a greater understanding of how to care for plants as they complete their life cycles. Children apply what they have learned from commercial food producers, as well as their own knowledge, to protect plants from weed growth and animal damage, and to ensure that they regularly provide plants with sufficient water and additional nutrients to encourage greater growth.

National curriculum links:

Describe the life process of reproduction in some plants

Learning intention:

To investigate ways in which crop yield might be improved

Scientific enquiry type:

Carrying out comparative and fair tests

Working scientifically links:

Planning different types of scientific enquiries to answer questions, including recognising and controlling variables where necessary

Success criteria:

- I can plan a scientific enquiry to answer a question about how to improve crop yield.
- I can recognise the need for a 'control' plant/s.
- I can identify what evidence I am going to collect throughout the investigation and suggest how this will help me to answer my initial question.

Key information:

This series of lessons should be linked closely with the Lesson 3 sessions, in which children are producing different crops to sell and eat. They can choose one of those crops as the focus of this investigation or, if more suitable or relevant, set up an investigation with new plants or seeds.

Key vocabulary:

crop, produce, yield, names of fruit and vegetables being grown

Resources:

A space with good soil for growing (or large containers and grow bags as an alternative), garden tools, compost or manure, seeds (and their packets) for salad crops (for example, lettuce, spring onion, rocket, nasturtium, pot marigold, cherry tomato, cucumbers, peppers) and flowering plants, potato tubers, strawberry plants, courgette or marrow plants

Health and safety:

Wash hands after working with soil or potting compost or handling plant material. Some plants may have sap that will irritate a child's skin. Some bulbs (for example, hyacinths) can also cause skin irritation.

EXPLORE:

Ask: *What might make a difference to how many crops the plants in our growing space will yield?*

Show children the video they saw for the first time at the end of Lesson 3 (Market garden; Video 1).

Ask: *What did the market gardener say that they did to avoid insect damage? How do they ensure that plants grow well and produce the maximum amount of crops?*

Ensure that children mention the use of plants, such as marigolds to distract insects, and refer to where they position plants. They should recognise the importance of regular weeding, hoeing, watering and feeding to maximise plant growth.

ENQUIRE:

Explain to children that during this series of lessons they are going to choose and investigate one method of improving crop yields. Encourage them to brainstorm question ideas, using the different variables that might affect growth – insect damage, weed growth, watering, feeding (using plant food), the distance between plants – as a starting point. Explain to them that during the initial lesson they are going to plan an investigation to test their method of improving crop yield and answer their question.

Challenge 1: Children complete an investigation framework to plan a comparative test

Provide the children with the Example planner (Resource sheet 1) to help them to plan their investigation. Advise them that the completed planner should be used as a guide and that they should come up with their own idea to test if possible.

Ask: *How will you know whether what you are doing is making a difference to how well the plants grow? How will you decide how well the plants have grown? What will you observe or measure that might provide you with evidence?*

Challenge 2: Children plan a comparative test

Ask the children to use their Our Changing World diaries to record their plan for their investigation. Provide them with the Investigation checklist (Resource sheet 2) to help them to ensure that they plan their investigation systematically.

Challenge 3: Children plan an investigation to identify an organic method to prevent crop damage by animals

Explain to the children that a local gardener has reported a problem at his allotment. His strawberries are being eaten by slugs or birds and some of his vegetable plants are being ruined by carrot fly. He only uses organic methods in his garden. What can he do to ensure that he has something to enter in the local produce show in September?

Explain to the children that their challenge is to plan and carry out an investigation that will provide evidence that they can use to suggest things that the gardener could do to improve the situation.

REFLECT AND REVIEW:

Encourage children to share their plans for their investigations.

Ask: *What are you trying to find out? What question are you investigating? How are you going to set up your enquiry? What will you look for over time? What evidence might you collect? What should you see if what you are doing to your crop is making a difference? How will you know that it is your investigation that has helped the crop to grow better?*

Remind children to include a control plant or plants in their investigations. The control area of planting is left undisturbed so that the plants here can be compared with the plants in the area that is the subject of the investigation.

Explain to children that during the next lessons in this series they will begin growing their plants (or set up their tests using plants that are already growing), and then continue their enquiry over an extended period of time (at least 6–8 weeks). At the end of the sequence of lessons. towards the end of summer term, they should consider the evidence that they have gathered, make comparisons between methods and discuss which method appears to have had the greatest effect on crop yield. Children taking on Challenge 3 will need to think about what evidence they can provide to the local gardener.

EVIDENCE OF LEARNING:

Listen to children as they share ideas during the Explore part of the session. Are they able to identify a variable to test based on these ideas? How effectively do they plan for their enquiry? Do they recognise the need for a control plant or plants that are not being well cared for? Can children identify what they should see if their crop is growing better over time? Can they identify what evidence they need to collect and suggest how this will help them to answer their initial question?

CIRCLE OF LIFE

INTRODUCTION

In this module children build on earlier work from Key Stage 1 and from Year 3, where they learned about the life cycles of plants. They extend their understanding of what a life cycle is, and learn about the life cycles of some familiar (and some less familiar) mammals, amphibians, insects and birds. Children compare and contrast different life cycles, identifying common features as well as explaining key differences. They use their knowledge of life cycles to help them as they create a fantastical creature of their own, complete with its own distinct life cycle. They learn about incredible journeys that some animals undertake to complete their life cycles, and about the different ways in which humans are supporting some endangered animals to increase their population numbers.

When working scientifically during this module, children frequently use secondary sources of information, as they carry out their own investigations to answer a variety of science questions, with increasing independence. This should involve the use of quality non-fiction books, web-based material, Apps, etc. and might include a visit to a local zoo, wildlife park or animal collection to gather information more directly from recognised experts.

Children report and present findings from their enquiries in a variety of ways, both orally and in written forms, drawing conclusions, identifying causal relationships and explaining their thinking. They consider evidence that has been used to support arguments, for example, as they learn about the work that has been done to protect the future of endangered animals.

This module links closely to other Year 5 modules, including Module 2: Reproduction in Plants and Animals, and should be taught before Module 2. It also links to OCW. Lessons in OCW provide children with opportunities to investigate and enquire practically into many aspects of the learning that is the focus of this module.

Key vocabulary:

life cycle, birth, growth, reproduction, metamorphosis, aging, death, animal, mammal, amphibian, insect, bird, elephant, toad, bumblebee, blue tit, hedgehog, bat, polar bear, mountain gorilla, cubs, pups, hibernate, nocturnal, marsupial, toad, newt, salamander, tree frog, metamorphosis, tadpole, larva, frog, toad, gills, cold blooded, ladybird, butterfly, dragonfly, head, thorax, abdomen, antennae, egg, pupa, cocoon, adult, thrush, peregrine falcon, ostrich, emperor penguin, breeding cycle, clutch, brood, hatch, fledge, prey, predator, reproduce, habitat, environment, humpback whale, blue whale, swift, osprey, wildebeest, caribou, monarch butterfly, migrate, migration, navigate, genetic, endangered, threatened, extinct, extinction, evolution, giant panda, black rhino, peregrine falcon, bumblebee, salamander, osprey, koala bear

FACT FILE:

Definition of an animal

An animal is any living thing that can move from place to place independently and has senses that help it to recognise and react to the world around it. Animals are unable to make their own food and so have to feed on other living things.

Animal life cycles

A life cycle is made up of a series of developmental changes that an organism goes through, as they are born, grow, develop to adulthood, reproduce, reach old age and die. The stages of the life cycle and length of that cycle vary, depending on the type of animal.

Mammals

Mammal life cycles vary significantly in length. They give birth to live young which look like smaller versions of the adult animal. Dogs, cows, elephants, mice, whales and humans are all mammals.

Amphibians

Amphibians spend part of their life cycle in water and part of their life cycle on land. Amphibians hatch in water from jelly-like eggs and, during their early stages of life, breathe using gills. They then develop lungs so the adults can breathe air and live on land. This process, which changes their appearance entirely, is known as metamorphosis. Frogs, toads and salamanders are all amphibians.

Insects

The life cycles of insects vary, but most insects hatch from eggs. The immature stages can be very different from adults in structure, habit and habitat. Those insects that go through complete metamorphosis have a pupal stage. Insects that go through incomplete metamorphosis do not have a pupal stage, but develop into adults through a series of nymphal stages instead. Bees, ants, butterflies, beetles and flies are kinds of insects.

Birds

Birds lay eggs that have hard shells. These eggs hatch out after a period of incubation. Young chicks are largely helpless and are fed by the adult birds until they have grown and developed sufficiently to leave the nest. Exceptions include duck chicks, which can run around almost immediately and are brought to water by their parents. Robins, eagles, chickens and ostriches are a few of the many kinds of birds.

Migration

Many animals migrate. The distances they travel vary hugely. Amphibians travel the shortest distances. They tend to migrate within their local area, to return to the water courses from where they originated, often negotiating busy roads en masse. Insects and birds travel great distances, but there are many incredible journeys undertaken by many animals as part of their life cycles.

Threatened and endangered animals

At any one time there are many species of animals that are considered to be threatened or endangered. The picture is ever changing and a source of information on endangered species, such as the International Union for the Conservation of Nature (IUCN) 'Red List', is reliable and updated regularly. Approximate figures are listed below.

Around 1 in 5 mammals are considered to be endangered. There are 5500 species of mammal and 1140 of these are endangered.

Around 1 in 8 birds are endangered. There are 10 064 species of bird and 1313 of these are endangered.

Around 1 in 3 amphibians are endangered. There are 6771 species of amphibian and 1931 of these are endangered.

Given that there are an estimated one million species of insect (about half the total number of living organisms), only a tiny proportion of these – about 4230 species – have been identified as threatened.

Common misconceptions:

Children may think that humans are not animals.

Children tend only to recognise common mammals as animals and do not include birds, insects, fish and amphibians.

Children may not appreciate that different types of animals have different life cycles, for example, they may think that all young animals start life as miniatures of their adult parents.

Big Cat book links

Kings of the Wild Jonathan and Angela Scott 978-0-00-723085-3 Band 13 Topaz	BBC documentary makers take us deep into the Kingdom of the brown bear.
Jaws and Claws and things with Wings Valerie Bloom 978-0-00-746539-1 Band 14 Ruby	A poetry collection inspired by the strange and wonderful creatures in the natural world.
On Safari Jonathan and Angela Scott 978-0-00-723125-6 Band 15 Emerald	Find out what it is really like to go on safari.
The Leopard Poachers Kathy Hoopman 978-0-00-733639-5 Band 16 Sapphire	Sameer and Ali escape through the Arabian mountains, terrified that they will fall victim to leopard poachers.
Life Cycles Sally Morgan 978-0-00-733640-1 Band 16 Sapphire	Explore the fascinating life cycle of the Salmon in this highly photographic book.

CIRCLE OF LIFE

LESSON 1: WHAT IS A LIFE CYCLE?

LESSON SUMMARY:

In this lesson children build on previous learning from Year 2 about the process of growth and reproduction in animals, and from Year 3 about the life cycles of flowering plants. They are introduced to the life cycles of four significant types of animals: mammals, amphibians, insects and birds. By the end of the lesson children have a deeper understanding of the main stages of an animal life cycle: birth, growth, reproduction, aging and death. They compare and contrast different animal life cycles, identifying common features and differences.

Key vocabulary:

life cycle, birth, growth, reproduction, metamorphosis, aging, death, animal, mammal, amphibian, insect, bird, elephant, toad, bumblebee, blue tit

Resources:

Mini whiteboards, sticky notes, secondary sources for research, including quality non-fiction books, web-based resources, educational CDs, smartphone and tablet Apps, identification guides and leaflets

National curriculum links:

Explain the differences in the life cycles of a mammal, an amphibian, an insect and a bird

Learning intention:

To compare the life cycles of different animals

Scientific enquiry type:

Using a wide range of secondary sources of information

Working scientifically links:

Reporting and presenting findings from enquiries, including conclusions, causal relationships and explanations of and degree of trust in results, in oral and written forms such as displays and other presentations

Success criteria:

- I can define the main stages of an animal life cycle.
- I can identify similarities and differences between the life cycles of an elephant, a toad, a bumblebee and a blue tit.
- I can describe how the length of life cycle of these animals varies.

Key information:

An animal is any living thing that can move from place to place independently and has senses that help it to recognise and react to the world around it. Children may think that humans are not animals. Children tend only to recognise common mammals as animals and may not include birds, insects, fish and amphibians. They may not appreciate that different types of animals have different life cycles, for example, they may think that all young animals start life as miniatures of the adults.

EXPLORE:

Show children a set of Plant life cycle sequence cards (Resource sheet 1), one at a time and in no particular order. The cards fit together to make up the life cycle of a runner bean.

Ask: *What do these pictures show? What is each stage called?*

Next, show children the series of images in the Life cycle slideshow (Slideshow 1), which illustrates an animal life cycle.

Ask: *What is a life cycle? What stages is it made up of? Can you come up with a sentence that describes what a life cycle is?*

Ask children to work with a partner to record their ideas on a mini whiteboard. After a few minutes, take feedback and note children's responses on the interactive whiteboard (or on a flip chart). Summarise their ideas as you talk with them and ensure that a clear, common sequence is identified: birth, growth, reproduction, aging and death.

ENQUIRE:

Explain that today's lesson focuses on the life cycles of some important types of animals. Check that they are clear about the meaning of the term 'animal'.

During this lesson children compare the life cycles of an elephant, a toad, a bumblebee and a blue tit, and identify common features and noticeable differences.

In Challenge 1, children order the stages of the life cycles of the four animals and identify significant differences between those life cycles. In Challenge 2, children order life cycle stages and use secondary resources to help them answer a question about those life cycles, while in Challenge 3 children compare life cycles.

The challenges are presented on the Challenge slides to be displayed on the board, or printed out and placed in the centre of the table.

Challenge 1: Children work in pairs and sequence a set of cut out Life cycle cards (Resource sheet 2)

Give each pair of children a set of the cut out Animal life cycle cards, which contain words or sentences describing the four different stages (birth, reproduction, aging and death) of the four different animals.

Ask: *Can you put the cards of the main life cycle stages into the correct order? Do you need to add any additional blank cards into the sequence? Do you notice any differences between the life cycles of different animals?*

The children should record the outcomes of their sort, ideally by photographing the sequence, uploading and annotating on screen or using sticky notes to annotate printed photographs.

Ask: *Which life cycle or cycles produce live young that look very like their parents, only smaller? Which don't? Which animals look very different at certain points in their life cycle?*

Encourage the children to talk in their pairs about the different stages of each animal's life cycle, and their similarities and differences.

Challenge 2: Children work in pairs to sequence a set of life cycle cards and use secondary sources of information to find out more about the differences between their life cycles

Give each pair of children a cut out set of the Animal life cycle cards (Resource sheet 2) and ask them to put the cards into the correct order, relating them to the main life cycle stages. Ask the children whether they need to add any additional blank cards into the sequence to show any missing stages of the life cycle. Do they notice any differences between the life cycles of the different animals?

Ask: *What type of animal is an elephant, a toad, a bumblebee or a blue tit? How are the life cycles of these animals different? What do they have in common?*

The children should use secondary sources of information, particularly suitable non-fiction books, but also perhaps web-based resources, CDs, smartphone and tablet Apps, identification guides and leaflets, to help them answer the questions.

Challenge 3: Children compare the life cycles of the four animals and use secondary sources to discover the average length of their life cycles

Begin with a focus question: Do all life cycles last for the same length of time? Ask the children to use the available information sources and draft a table or make notes to record the differences they discover.

Ask: *How long before a female can reproduce? How many young might a female have? At one birth? Over many years? How long might the animal live?*

Prompt the children to think about whether the number of offspring an animal has is related to its own life span.

REFLECT AND REVIEW:

Ask children to share what they have learned about the life cycles of the animals they have explored. Children who completed Challenges 1 and 2 could show their completed life cycle sequences. Encourage children to describe the sequences, using appropriate vocabulary, and to describe differences that they notice between the life cycles.

Show children a set of the Turtle life cycle cards (Resource sheet 3). Again, stick the cards into position in response to children's directions. Ask them which of the other animal life cycles is the most similar to this reptile's life cycle.

Ask: *How long does a life cycle last? What does it depend on?*

Encourage children who completed Challenge 3 to provide detail based on their research from secondary sources.

Ask: *Which of the animals you have investigated has the longest average life cycle? Which has the shortest? Do you think that the life spans of other types of mammals would be as long as an elephant's?*

EVIDENCE OF LEARNING:

Listen to children as they develop their understanding of the life cycles of the four different animals, giving examples of a mammal, an amphibian, an insect and a bird. Can they define what a life cycle is? Can they describe the main stages of an animal life cycle? Do they identify common features and differences in the life cycles of the elephant, a toad, a bumblebee and a blue tit? Can they name the group of animals to which each belongs? Can they explain variation in the length of life cycles between these animal types?

CROSS-CURRICULAR OPPORTUNITIES:

This lesson can be linked to English with reading comprehension and non-fiction texts.

CIRCLE OF LIFE

LESSON 2: WHAT DO WE KNOW ABOUT THE LIFE CYCLES OF MAMMALS?

Key vocabulary:

life cycle, hedgehog, bat, polar bear, mountain gorilla, cubs, pups, hibernate, nocturnal, marsupial

Resources:

Mini whiteboards, secondary sources for research, including quality non-fiction books, web-based resources, educational CDs, smartphone and tablet Apps, identification guides and leaflets

Key information:

Mammal mothers give birth to live young and produce milk to feed their babies. Most mammals have four legs or two arms and two legs, but not all mammals live on land: mammals that spend their whole life cycle in aquatic environments have developed fins and flippers. Dogs, cows, elephants, mice, whales and humans are all mammals.

Key information:

Mammals have been selected – hedgehog and Pipistrelle bat because they should be relatively familiar British mammals, while the polar bear and mountain gorilla are mammals from further afield that children may have some knowledge about.

LESSON SUMMARY:

In this lesson children deepen their knowledge from Year 3 about the group of animals called mammals. By the end of the lesson they will have found out about the life cycles of a variety of mammals, identifying some common characteristics.

National curriculum links:

Explain the differences in the life cycles of a mammal, an amphibian, an insect and a bird

Learning intention:

To define what a mammal is and describe its life cycle

Scientific enquiry type:

Using a wide range of secondary sources of information

Working scientifically links:

Reporting and presenting findings from enquiries, including conclusions, causal relationships and explanations of and degree of trust in results, in oral and written forms such as displays and other presentations

Success criteria:

- I can define what a mammal is.
- I can describe the common characteristics of different types of mammal.
- I can sequence the life cycle stages of a hedgehog, a bat, a polar bear or a mountain gorilla.
- I can ask questions to find out more about mammals and identify how to answer these questions.

EXPLORE:

Display the first part of Different life cycles? (Interactive 1). Show children the images of the four animals that they learned about during the last lesson: toad, bumblebee, blue tit and elephant. Check whether children can remember the name and type of each animal by looking at and discussing them in sequence and then matching them to their correct type on screen.

Ask: *What is this animal called? Which type of animal is it? What did you learn about its life cycle during the last lesson?*

Then move on to the second part of Interactive 1.

Ask: *Are their young born live? How do they change as they grow? Which of the animals changes the most during its life cycle? How do they reproduce when they become adults? How long do they live?*

Drag and drop answers into the boxes on screen. Ensure that children are beginning to build up a clearer picture of the differences between the life cycles of the different types of animals. This understanding will develop further and become more detailed over the course of the next four lessons, as children explore the four types of animals more closely.

ENQUIRE:

Explain to children that during this lesson they are going to learn about the life cycle of some mammals in much greater detail.

Show children Marvellous mammals (Video 1), which features the four mammals that they are going to find out about during the lesson: hedgehog, bat, polar bear and mountain gorilla.

Ask children whether they know the names of any or all of the four animals. Stick Mug shots (use Set 1 from Resource sheet 1) of each of the mammals, and their names, where children can see them.

In Challenges 1 and 2 children use secondary sources to find out more about either a hedgehog, the Pipistrelle bat, a polar bear or a mountain gorilla. They are provided with a differentiated resource asset on which to record the evidence that they collect. In Challenge 3, children use secondary sources to find out about the life cycle of a mammal of their choice, from a selection, and use the information to develop a 'Whose life cycle is it anyway?' game.

The challenges are presented on the Challenge slides to be displayed on the board, or printed out and placed in the centre of the table.

Challenge 1: Children work in pairs and use secondary sources to find out information about the life cycle of one of the mammals

Give children a copy of Life cycle table (Resource sheet 2), which asks key questions that they should research using secondary sources and on which they should record their answers.

Challenge 2: Children work in pairs and use secondary sources to find out information about the life cycle of one of the mammals, using the KWLH table (Resource sheet 3) to record their research

Give the children in pairs the KWLH Life cycle grid, which asks key questions that they should research using secondary sources. They should record their answers on the grid.

Challenge 3: Children work in pairs to research the life cycle of a mammal and use the information they discover to develop a 'Whose life cycle is it anyway?' game

Show the children Set 2 of the Mug shots from Resource sheet 1 and ask each pair of children to choose an animal to research. Ask the children to use the information they collect through their research to develop a 'Whose life cycle is it anyway?' game. They will need to write statements that are correct (but that don't make the mammal too easy to identify). The aim is to say as many true facts as possible about the life cycle of the mammal, before the class can guess what it is. The quicker the class identifies the mammal, the more points they gain.

REFLECT AND REVIEW:

Encourage some children who have completed Challenge 3 to each lead a round of the 'Whose life cycle is it anyway?' game.

Ask children to find a partner to work with who has researched a different mammal from them and together to complete a Compare and contrast grid (Resource sheet 4).

Explain to children that you are going to introduce them to a mammal with some different stages in its life cycle.

Show an image of a kangaroo (Slideshow 1). Explain that a kangaroo is a special type of mammal called a marsupial. Ask them whether they know anything about kangaroos already. If a child mentions the pouch, ask them what the pouch is for. If necessary, explain to children that marsupial young are born long before they can survive away from their mother and that they spend anything up to six months in their mother's pouch drinking her milk, until they are developed enough to come out. Children might like to find out more about marsupials as part of a homework project.

EVIDENCE OF LEARNING:

Listen to children as they carry out research and reflect on what they have learned about the life cycles of mammals. Look at their written work. Can they define a mammal? Can they describe the common characteristics of mammals? Can they describe the life cycle of the mammal that they have studied in detail, referring to each stage of the life cycle? Did children frame appropriate questions to find out more about mammals? Can they explain how the life cycle of a marsupial is different from that of other mammal types?

CROSS-CURRICULAR OPPORTUNITIES:

This lesson can be linked to English, through reading comprehension, research skills, and non-chronological writing.

Key information:

At birth, marsupial babies are not fully developed. The baby's hind legs are just nubs. The baby lives and continues to develop in the mother's pouch until it is developed enough to move around independently. The pouch, or marsupium, is also where the mother's mammary glands are, for feeding the baby. A baby kangaroo may live in its mother's pouch for six months.

CIRCLE OF LIFE

LESSON 3: WHAT DO WE KNOW ABOUT THE LIFE CYCLES OF AMPHIBIANS?

LESSON SUMMARY:

In this lesson children deepen their knowledge from Year 3 about the group of animals called amphibians. By the end of the lesson they will have found out about the life cycles of a variety of amphibians, identifying some common characteristics including the process of metamorphosis.

Key vocabulary:

life cycle, amphibian, toad, newt, salamander, tree frog, metamorphosis, gills, cold blooded

Resources:

Mini whiteboards, secondary sources for research, such as quality non-fiction books, web-based resources, educational CDs, smartphone and table Apps, identification guides and leaflets

National curriculum links:

Explain the differences in the life cycles of a mammal, an amphibian, an insect and a bird

Learning intention:

To define an amphibian and describe its life cycle

Scientific enquiry type:

Using a wide range of secondary sources of information

Working scientifically links:

Reporting and presenting findings from enquiries, including conclusions, causal relationships and explanations of and degree of trust in results, in oral and written forms such as displays and other presentations

Success criteria:

- I can define what an amphibian is.
- I can describe the common characteristics of different types of amphibian.
- I can sequence the life cycle stages of a toad, newt, salamander or tree frog.
- I can decide what information sources to use to find out about amphibians.
- I can explain what is different about the life cycle of an amphibian compared with that of a mammal.

EXPLORE:

Show children Hedgehog and garden frog (Video 1) and then use the Compare and contrast grid (Resource sheet 1) to compare the lives and life cycles of these two types of animals.

Through discussion, help children to recognise information about the two animals that is directly relevant to the life cycles of each. Some of the facts they may suggest, while interesting, may not be of direct relevance to life cycles.

Ask: *Where do hedgehogs live? Do they live in the same place for all the seasons of the year? Where do hedgehogs build their nests and give birth to their young? Is this information important to the hedgehog's life cycle?*

Children might argue that the home and habitat of the hedgehog is relevant, as the animal will fail to reproduce if it cannot access an appropriate territory with the right sort of food and cover.

Ask: *What does a frog eat? Is this relevant to our work on the life cycles of amphibians?*

Again, children might argue that food supply is important to the life cycle of the frog. It is unlikely that frogs will lay eggs in a pond or water course where food for tadpoles is limited.

ENQUIRE:

Explain to children that during this lesson they are going to learn more about the life cycle of an amphibian. Look again at the frog's life cycle in Video 1. Ask them to work in pairs to summarise the frog's life cycle in a description or diagram, using their mini whiteboards to record their ideas.

The challenges are presented on the Challenge slides to be displayed on the board, or printed out and placed in the centre of the table.

Challenge 1: Children become experts on the life cycle of a toad

Explain to children that they are going to become experts on the life cycle of a particular amphibian – the toad. Give each child or pair a key question to research from the following: Where do toads live during the course of their life cycle? How do they reproduce? How does their appearance change as they grow? How do they breathe at different stages of the life cycle?

Key information:

Amphibians are small animals that spend part of their life cycle in water and part of their life cycle on land. Amphibians hatch in water and during the early part of their life cycle breathe with gills. They then develop lungs so that the adults can breathe air. This process, which changes their appearance entirely, is known as metamorphosis. Amphibians are cold-blooded animals with skeletons inside their bodies. Frogs, toads and salamanders are amphibians.

Key information:

'Jigsawing' is a collaborative teaching and learning strategy. Children become 'expert' about a certain aspect of a topic and then teach others what they have learned. In this case, children have a particular question to investigate as a focus for their research. Once they have completed their task they share what they have learned to become toad 'experts'.

Ask the children to use secondary sources, such as books and online resources, to find out as much as they can to answer their question and to make notes on Fabulous facts about toads (Resource sheet 2), identifying the facts that they have discovered and recording the sources of information they have used. Then ask them to work together as a group to 'jigsaw' the information, combining everybody's research so that they can be considered experts on toads.

Challenge 2: Children become experts on the life cycle of a salamander

Explain to the children that they are going to work together to become salamander experts. Give each child or pair a question to research from the following: Where do salamanders live? How do they change in appearance during their life cycles? How do they breathe? What do they eat? How long do they live?

Ask the children to find out as much as they can about their question in 5 minutes, identifying the sources of information they use as they go. They then 'jigsaw' the information they have noted down and combine everybody's research so that they can be considered experts on salamanders.

The children should then produce an Expert's crib card (Resource sheet 3), summarising the key facts that they have learned about the life cycle of a salamander.

Challenge 3: Children use secondary sources to investigate the variety of amphibian life around the world

Explain to the children that they are going to find out about some amazing amphibians that exist across the world. They should use their own preferred form of recording to note key pieces of information about the life cycles of these amphibians and record sources of their information.

Ask: *How many amphibians have you been able to identify? Can you be sure that the information you've recorded is reliable? Do you trust the sources you have used? Where do these amphibians live? How do their life cycles compare? Have you found any examples of amphibians that seem to be different from most of the others?*

Challenge the children's thinking by bringing one particular amphibian into the discussion.

Ask: *Did you discover an unusual amphibian called a mudpuppy? What's different about it? How does the mudpuppy's life cycle compare with the life cycles of other amphibians?*

Encourage the children to work together to discover the answers to the mudpuppy questions.

REFLECT AND REVIEW:

Encourage children to share their growing expertise about the specific amphibians that they have researched.

Ask: *How are the life cycles of the amphibians you've learned about during this lesson similar? What differences have you noticed?*

Children who completed Challenges 1 or 2 should know that the amphibians that they have become expert about live for some part of their life cycle in water – where they lay their eggs – and part of their lives on land. Children who completed Challenge 3 should be able to explain that some amphibians instead spend their whole life cycle in the water.

Refer children back to the Compare and contrast grid that they worked on during the Explore part of the lesson (Resource sheet 1).

Ask: *What is the biggest difference between the life cycle of mammals and amphibians? What does a baby hedgehog (hoglet) look like when compared with an adult? How about a tadpole? Does it look like the adult frog? Why not?*

The major concept here is metamorphosis. Hedgehog young are just smaller versions of their parents, while tadpoles bear no resemblance to frogs! Amphibians all undergo significant changes in appearance and form during metamorphosis. During the next lesson of this module, children learn about metamorphosis in insects.

Key information:

The mudpuppy and axolotl are unusual types of salamander. They both live their whole lives in water, retaining feathery gills towards the backs of their heads, which they use to breathe. In this respect these two salamanders are different to most types of amphibian, which spend the early part of their lives in water, where they breathe using gills, and the latter part on land, where they breathe using lungs that develop as they grow towards adult stage.

EVIDENCE OF LEARNING:

Listen to children as they carry out research and reflect on what they have learned about the life cycles of amphibians. Look at their fact cards. Can they define an amphibian? Can they describe the common characteristics of amphibians? Do they recognise that amphibians go through metamorphosis as part of their life cycle? Can they describe the life cycle of the amphibian they have studied in detail, referring to each stage? Can they select appropriate sources of information? Can they explain the main similarities and differences between the life cycle of an amphibian and a mammal?

CIRCLE OF LIFE

LESSON 4: WHAT DO WE KNOW ABOUT THE LIFE CYCLES OF INSECTS?

LESSON SUMMARY:

In this lesson children deepen their knowledge from Year 3 about the group of animals called insects. By the end of the lesson they will have found out about the life cycles of a variety of insects, identifying some common characteristics including metamorphosis.

National curriculum links:

Explain the differences in the life cycles of a mammal, an amphibian, an insect and a bird

Learning intention:

To define what insects are and describe the different types of life cycle, including the process of metamorphosis

Scientific enquiry type:

Using a wide range of secondary sources of information

Working scientifically links:

Reporting and presenting findings from enquiries, including conclusions, causal relationships and explanations of and degree of trust in results, in oral and written forms such as displays and other presentations

Success criteria:

- I can define insects.
- I can describe the common characteristics of different types of insect.
- I can sequence the life cycle stages of an insect.
- I can present my findings in a poster.
- I can describe the differences and similarities between metamorphosis in insect life cycles and an amphibian life cycle.

Key vocabulary:

life cycle, insect, species, ladybird, butterfly, beetle, dragonfly, head, thorax, abdomen, antennae, metamorphosis, egg, larva, pupa, cocoon, adult

Resources:

Secondary sources for research, including quality non-fiction books, web-based resources, CDs, smartphone and tablet Apps, identification guides and leaflets

Key information:

Insects are small animals with a hard covering over their bodies, which is known as an exoskeleton (this is a skeleton, but it is on the outside in their bodies). Insects have a body that is divided into three parts: head, thorax and abdomen. Most insects also have three pairs of legs and one or two pairs of wings. They have a pair of feelers on their heads, called antennae. There are two versions of an insect life cycle. One is referred to as complete metamorphosis and involves a pupal stage. The other is incomplete metamorphosis and does not involve a pupal stage. Instead, in incomplete metamorphosis growth and development occur through a series of nymphal stages, and the insect sheds its skin a number of times.

EXPLORE:

Ask: *What is an insect?*

Ask children (give them 2 minutes) to draw a sketch and label an insect of their choice, which shows in as much detail as possible what they already know. After the 2 minutes have passed, allow them to look at other children's drawings and give them another 2 minutes to 'magpie' any good ideas that they see.

Show children Insects galore! (Video 1), reminding them to watch carefully and to remember any additional information that they could add later to their insect sketch.

Ask: *What did you learn that you didn't know already? What additional information do you need to add to your insect sketch? Do we know enough now to come up with a definition of an insect?*

Encourage children to talk to a partner before collecting their ideas. Ensure that they pick up on the key points and agree a definition.

ENQUIRE:

Remind children that in Lesson 3 they learned about a process that amphibians go through as they change from egg to adult.

Ask: *What is this process called?*

Explain that they are going to spend some time during this lesson finding out about some more 'masters of metamorphosis' but this time they will be learning about insects.

Children work in groups, using secondary resources to research an insect and create a poster explaining its life cycle. Children then decide which of the insects goes through the most amazing example of metamorphosis during its life cycle.

The challenges are differentiated according to the amount of support that is provided to children to present their findings and through the nature of the task.

Bees, ants, butterflies, ladybirds, beetles, moths, houseflies, wasps, caddisflies and mosquitoes are examples of insects that go through complete metamorphosis. Damselflies, dragonflies, crickets, stick insects, water boatman, cockroaches, mayflies and locusts are examples of insects that go through incomplete metamorphosis, which involves a series of nymphal stages, either in water or on land.

Challenge 1: Children produce a life cycle poster for either a ladybird, butterfly, beetle or dragonfly and record in words and pictures what they have learned through their research

Provide the children with an Outline life cycle poster (Resource sheet 1) to complete. They need to use secondary sources of reference, including quality non-fiction books, to help them identify key information about the life cycle of one of four insects: ladybird, butterfly, beetle or dragonfly. They should record as much information as they can about their 'master of metamorphosis' on their poster.

Ask: *What insect did you find out about? What is its 'species' name (for example, red admiral)? What stages does it have in its life cycle? How much does it change as it goes through the process of metamorphosis?*

Challenge 2: Children plan and produce a poster showing the life cycle of an insect of their choice, using secondary resources to help them select key information to use in their poster

Ask the children to decide on an insect that will be the focus for their poster. If necessary provide the List of insects sheet (Resource sheet 2) for them to choose from. Provide the children with secondary sources to aid in their research and explain that they should use the Poster planner (Resource sheet 3) to organise the information that they collect. The poster should communicate information about all the stages of the insect's life cycle and the children should record as much information as they can about their 'master of metamorphosis'.

Ask: *What is the species name of your insect? What did you find out about its life cycle? How many stages does your insect's life cycle have? How long does each stage of metamorphosis last? Does it have a larval stage or not?*

Challenge 3: Children choose a method for communicating the life cycle of their chosen insect

Encourage the children to research their chosen insect using a variety of reference sources, identifying the stages of its life cycle and recording as much information as they can about their 'master of metamorphosis'. The children might choose to combine images and information (on screen or as sound clips) in a digital presentation.

Ask: *What is the species name of your insect? What did you find out about its life cycle? How long does each stage of its metamorphosis last? Does it have larval and pupal stages? If so, what happens to the insect at the pupal stage? If not, does the life cycle include a nymph stage? If so, where does the nymph live? What happens to the nymph as it grows?*

If the children have chosen an insect with a larval stage, for example, a butterfly, moth, bee or housefly, suggest that they look at a second example that has nymph stages instead, for example, a water boatman, dragonfly, ladybird, stick insect or cricket. Encourage them to compare and contrast the different stages of metamorphosis for each group of insects, recognising that one (complete metamorphosis) includes a 'rest' stage while the insect is in a pupa.

REFLECT AND REVIEW:

Ask children to share what they have found out about their 'masters of metamorphosis'. Can they identify two distinct types of metamorphosis? Help them to divide the insects they have found out about into those that go through a complete metamorphosis, which involves a pupal stage, and those that show incomplete metamorphosis and develop through a series of nymphal stages. Which changes are the most dramatic?

Children should then vote for the insect that they feel goes through the most dramatic or awe-inspiring changes as part of its life cycle.

EVIDENCE OF LEARNING:

Listen to children as they carry out research and reflect on what they have learned about the life cycles of insects. Look at their posters. Can they define an insect? Can they describe the common characteristics of insects? Do children recognise that insects go through metamorphosis as a part of their life cycle? Can they describe in detail the life cycle of the insect that they have studied, referring to each stage of the life cycle? Do they present the information clearly in a poster, using labelled drawings and text?

CIRCLE OF LIFE

 LESSON 5: WHAT DO WE KNOW ABOUT THE LIFE CYCLES OF BIRDS?

LESSON SUMMARY:

In this lesson children deepen their knowledge from Year 3 about the group of animals called birds. By the end of the lesson they will have found out about the life cycles of a variety of birds, identifying some common characteristics.

Key vocabulary:

life cycle, bird, thrush, peregrine falcon, ostrich, emperor penguin, chicken, breeding cycle, brood, hatch, fledge

Resources:

Secondary sources for research, including quality non-fiction books, web-based resources, CDs, smartphone and tablet Apps, identification guides and leaflets

Key information:

Birds are animals with two wings, two legs and a body covered with feathers. Most birds can fly. Birds are warm-blooded animals with skeletons inside their bodies. Birds lay eggs with hard shells. Robins, eagles, chickens and ostriches are a few of the many different kinds of birds.

National curriculum links:

Explain the differences in the life cycles of a mammal, an amphibian, an insect and a bird

Learning intention:

To define what a bird is and describe its life cycle

Scientific enquiry type:

Using a wide range of secondary sources of information

Working scientifically links:

Reporting and presenting findings from enquiries, including conclusions, causal relationships and explanations of and degree of trust in results, in oral and written forms such as displays and other presentations

Success criteria:

- I can define a bird.
- I can describe the common characteristics of different types of birds.
- I can sequence the life cycle stages of a bird.
- I can select and record key information gathered from secondary sources to help me answer questions about the breeding cycle of birds.

EXPLORE:

Show children the Birds as animals video (Video 1). There are around 10 000 different species of birds in the world and almost 600 in Great Britain, but what do we know about them? Ask children to work with a partner to agree a definition of a bird. Remind them to include key characteristics.

Use Things that fly (Resource sheet 1) to challenge children's thinking. Ask them to consider how to answer this: A bird is like an aeroplane because … ? Children hopefully identify similarities, for example, they can fly, they have wings, they have 'landing gear' to help them land, they need fuel to fly, and they travel long distances very quickly. Children should then discuss the question with each other. Ask them to agree on some important differences between a bird and an aeroplane.

ENQUIRE:

Explain to children that they are going to explore and answer questions about the life cycles of birds, working in groups to look at bird life cycles, and share their findings.

All three challenges involve children using secondary sources of reference, with varying amounts of support, to answer key questions.

The challenges are presented on the Challenge slides to be displayed on the board, or printed out and placed in the centre of the table.

Challenge 1: Children identify the owner of one of four 'mystery eggs' and use secondary sources to find out about the breeding cycle of the bird that laid the egg

Provide the children with images of the eggs of four birds (Birds' eggs; Resource sheet 2): thrush, peregrine falcon, ostrich and chicken. Keep the producer of the eggs secret until later in the lesson.

The children's challenge is to use secondary sources to identify who laid each egg and to find out as much as possible about the breeding cycle of that particular bird. Each group should identify the owner of one egg.

Ask: *Whose egg could it be? Is it big or small? How many eggs are usually produced in each clutch? How many broods could parent birds produce each year? What could happen to a clutch of eggs or a nest full of chicks that might affect their survival?*

Ask the children to note down the key information using a table, so that they can tell others which egg is which later in the lesson.

Challenge 2: Children use secondary sources of information to find out about the life cycle of the peregrine falcon so they can answer specific questions about the bird's life cycle

Explain that sometimes in the past peregrine falcons have struggled to successfully raise chicks in their normal breeding locations. To help with this problem humans have created nest boxes on cathedrals all over Britain. Show the children information on the Hawk and Owl Trust website about the peregrine breeding support programme at Norwich Cathedral (http://upp.hawkandowl.org).

The children's challenge is to use the Peregrine falcon fact card (Resource sheet 3) and other secondary sources of information (including the Hawk and Owl Trust website) to help them answer these questions about the life cycle of the peregrine falcon: Where do peregrine falcons normally build their nests? How many eggs do they usually produce? Who looks after the chicks when they hatch? What food do the chicks eat? How often are they fed? How much do they need to eat each day? How long before the chicks are big enough to fledge and leave the nest? What happens to them next?

Remind the children to record key information that answers these questions, so that they can share their knowledge with others later in the lesson.

Challenge 3: Children use a wide variety of secondary sources to help them select relevant information about the breeding cycle of specific birds and to communicate this information in a plan for a 'Top Tip Guide for new bird parents'

Ask the children to imagine that they are a parent thrush, peregrine falcon, ostrich, or emperor penguin.

Ask: *What would you need to know about raising your new family?*

Ask the children, in pairs, to use secondary sources of information to help them to come up with a list of ideas for their Top Tip Guide for new bird parents. They should communicate what they find out in a 'Top Tips' plan for their guide.

REFLECT AND REVIEW:

Place the four Birds' eggs cards (Resource sheet 2) in a position where all children can see them.

Ask: *Where does the egg fit into the life cycle of a bird? Whose eggs might these be?*

Ask some children who identified the mystery eggs to share the evidence that they collected and recorded during the lesson. Which egg do they think belongs to which bird? If necessary, prompt children by asking key questions from the earlier challenge. The most important observations that can help with bird egg identification are the size of the eggs, the number of eggs in the clutch, the appearance, colour and pattern of the eggshells and – in a real life context – where the nest is located.

Children should identify that these are the eggs of a peregrine falcon, thrush, ostrich and chicken.

Ask: *What do we know about the breeding cycle of those birds? How are their breeding cycles different?*

If time allows, encourage children to share key information that they have learned about each of the four birds.

Some types of birds, for example, the peregrine falcon or emperor penguin, invest a lot of effort in only one (or two) eggs. Others, for example, the thrush and blue tit, produce larger numbers of eggs to increase the likelihood of some chicks surviving.

Explain to children that the raising of chicks is a vital part of the life cycle of a bird, but it is just part of that cycle. Ask them to consider and discuss with a partner what the other life cycle stages are.

Record the stages of the bird life cycle on a flip chart or whiteboard. Ensure that children recognise that when chicks fledge and leave the nest it is just the beginning of the story. Chicks surviving to adulthood breed and start the cycle again.

EVIDENCE OF LEARNING:

Listen to children as they carry out research and reflect on what they have learned about the life cycles of birds. Can they define a bird? Can they describe the common characteristics of birds? Can they describe in detail the life cycle of the birds that they have studied, referring to each stage of the life cycle? Do they select and record key information from secondary sources and use it to help them answer questions about the breeding cycle of birds?

CIRCLE OF LIFE

Key vocabulary:

life cycle, mammal, amphibian, insect, bird, prey, predator, reproduce, habitat, environment, metamorphosis, caterpillar, pupa, tadpole, butterfly, elephant, frog, mature, immature

Resources:

Secondary sources for research, including quality non-fiction books, web-based resources, CDs, smartphone and tablet Apps, identification guides and leaflets

LESSON 6: WHAT MAKES A SUCCESSFUL LIFE CYCLE?

LESSON SUMMARY:

In this lesson children apply their knowledge and understanding of animal life cycles to an unfamiliar context. By the end of the lesson they will have invented their own animal, described in detail each stage of its life cycle and explained how this will ensure its long-term success. Children will be able to answer questions about their animal, which will reveal how their understanding of life cycles has advanced.

In order to produce a quality outcome this lesson may be extended over several hours.

National curriculum links:

Explain the differences in the life cycles of a mammal, an amphibian, an insect and a bird

Learning intention:

To create a life cycle for an imaginary animal that will help to ensure its long-term success

Scientific enquiry type:

Finding things out using secondary sources of information

Working scientifically links:

Identifying scientific evidence that has been used to support or refute ideas or arguments

Success criteria:

- I can create a detailed life cycle for an imaginary animal.
- I can identify ways in which my animal will be successful.
- I can communicate my ideas creatively.

EXPLORE:

Provide each child with a Life cycle sticker (printed from Resource sheet 1) describing a single stage of a life cycle. Ask them to look at their sticker in secret before sticking it onto another child's back, without them seeing what it is. Each child then asks other children focused questions with 'yes, no or don't know' type answers, for example: Is it part of the life cycle of a mammal? Is it a stage of metamorphosis?, as they attempt to discover what the word or statement is. Once they have correctly guessed what's on their own sticker they should look for classmates who have stickers that might form part of the same life cycle.

ENQUIRE:

Explain that during this lesson they are going to create an imaginary animal and plan its life cycle in detail. Explain that they can communicate their ideas in a variety of ways, perhaps creating animations or other multimedia forms of presentation to depict each stage in the life cycle of their imaginary animal.

In Challenges 1 and 2 children are supported with a structured storyboard planner to use as they plan the stages of their imaginary animal's life cycle. Further prompting can help support children completing Challenge 1 as they decide how to communicate the stages of their animal's life cycle. In Challenges 2 and 3 teacher questioning explores children's understanding of the life cycles they are seeking to communicate and establishes whether children are able to justify the stages they have chosen to include.

The challenges are presented on the Challenge slides to be displayed on the board, or printed out and placed in the centre of the table.

Challenge 1: Children plan and communicate the life cycle of an imaginary animal, with support

Give the children working individually or in pairs the Storyboard planner (Resource sheet 2) to help them map out their animal's life cycle. Ask them to consider, with support if necessary, how best they can communicate the different stages of their animal's life cycle to the other children in the class.

Ask: *What is your imaginary animal called? What type of animal is it? What stages does your animal have in its life cycle? What part of the life cycle would you need to include if you were to convince somebody else that your animal is, for example, an insect or an amphibian?*

If necessary, provide suggestions or scaffolds to support the children as they decide how to communicate their ideas. They might, for example, create a poster using a life cycle frame, such as the one used in Lesson 4.

If possible, provide time for the children to use their chosen form of communication to complete their task.

Challenge 2: Children plan and communicate the life cycle of an imaginary animal, with support, and consider how the life cycle of their animal will ensure its success

Give the children working individually or in pairs the Storyboard planner (Resource sheet 2) to help them map out their animal's life cycle. They should decide how best to communicate the different stages of their animal's life cycle to other children in the class.

Ask: *What is your imaginary animal called? What type of animal is it? What stages does it have in its life cycle? How is your imaginary animal's life cycle different from that of other types of animals? What stage/s does your animal's life cycle have that would help someone know what type of animal it is?*

Encourage the children to extend their thinking further about their imaginary animal's ability to survive.

Ask: *How could your animal ensure that its life cycle is successful? How could it protect itself, at every stage of the life cycle? How will it make sure that both adults and young have enough to eat?*

Ask the children to think about the best way of sharing information about their animal and its life cycle with others. If possible, provide time for the children to use their chosen form of communication to complete their task.

Challenge 3: Children work individually or in pairs to plan and communicate the life cycle of their imaginary animal

Encourage the children to be as imaginative as possible in designing their animal and its life cycle. They might think about combining two types of animals to form a composite, including stages from the life cycle of a mammal and an insect, for example. If they do, then ask them what stages of each animal life cycle they will need to include, so that the other children will know which types of animals have been put together.

Ask: *How can your imaginary animal ensure that its life cycle is successful? How could it protect itself, at every stage of the life cycle? How will it make sure that both adults and young have enough to eat? What (if anything) might threaten the success of your animal?*

Give the children some time to think about the best way of sharing information about their animal and its life cycle with others, and, if possible, provide time for the children to use their chosen form of communication to complete their task.

REFLECT AND REVIEW:

Choose some children to present the life cycles of their imaginary animal to the class. Ask them to describe the type of animal it is and the stages in its life cycle. Children who completed Challenges 2 or 3 could be asked questions about their animal's survival and protection.

Explain that there are some real fantastical creatures in the natural world. Gradually click on the grid to reveal on screen a mugshot of a Duck-billed platypus (Interactive 1), starting with the 'beak' only (this will be on the left side of the grid) and continue to click on the grid to reveal the image until the whole animal can be seen.

Ask: *What is a duck-billed platypus? It lays eggs, so surely it must be a bird? What type of animal do you think it is, based on what we have learned already? How do you think it is able to protect itself against predators? What features does it have that resemble other animals, for example, tail like a beaver? Can you guess or work out what it eats?*

EVIDENCE OF LEARNING:

Listen to children as they go through the process of inventing their imaginary animal and its life cycle. Can they recall relevant subject knowledge and use it effectively? Can they explain how they have adapted the life cycle of their animal to make it more successful (for example, by ensuring that it can protect and feed itself at every stage of its life cycle)? Can they communicate their ideas creatively, ensuring that the information presented is clear, concise, scientifically accurate and includes appropriate vocabulary?

CIRCLE OF LIFE

LESSON 7: HOW ARE HUMANS HELPING ENDANGERED ANIMALS TO COMPLETE THEIR LIFE CYCLES?

LESSON SUMMARY:

In this lesson children find out about the ways in which humans are using science to help endangered animals complete their life cycles. By the end of the lesson they will know some of the reasons why different types of animals may become endangered and will also be able to argue a case for an animal under threat, suggesting why it should be helped and how science might be useful in doing this.

Key vocabulary:

endangered, threatened, extinct, extinction, evolution, giant panda, black rhino, peregrine falcon, bumblebee, salamander, osprey, koala bear

Resources:

Secondary sources for research, including quality non-fiction books, web-based resources, CDs, smartphone and tablet Apps, identification guides and leaflets, www.konicaminolta.com/kids/endangered_animals

National curriculum links:

Explain the differences in the life cycles of a mammal, an amphibian, an insect and a bird

Learning intention:

To explore and describe ways in which humans are using science to help endangered animals to complete their life cycles and increase their population numbers

Scientific enquiry type:

Finding things out using secondary sources of information

Working scientifically links:

Identifying scientific evidence that has been used to support or refute ideas or arguments

Success criteria:

- I can identify reasons why animals become endangered.
- I can describe ways in which humans are using science to help threatened or endangered animals.
- I can present evidence to support my ideas.

EXPLORE:

Display the beginning of the Decision alley slideshow (Slideshow 1). Children are presented with an initial statement: It's very important for humans to help animals that are endangered in the wild. Ensure that children understand the word 'endangered' and can relate this to their understanding of different types of animals and their life cycles.

Continue through the slides, which include pictures of endangered animals. A number of points of view are offered: 'The best way is to keep animals safe in zoos and help them breed there.' 'We should find ways of protecting their habitat, so that they have enough food to eat and can breed naturally.' 'If they die out, it's a shame but it doesn't really matter – it's all part of evolution.' 'Some animals are killed for food or to make medicines for people – is that wrong?' 'Breeding programmes cost a lot of money, helping people is more important.'

Ask: *What do you think?*

Encourage children to think, pair and share ideas with the rest of the class. Take feedback and note children's responses and ideas, as these will be revisited at the end of the lesson.

ENQUIRE:

Ask: *How do humans help endangered animals to reproduce, as part of their life cycle?*

Show children Endangered animals' stories (Video 1), which features the stories of two endangered mammals that British zoos are helping: the black rhino (breeding programme at Chester zoo) and the giant panda (breeding programme at Edinburgh zoo).

Link: Panda is artificially inseminated at Edinburgh Zoo

http://www.bbc.co.uk/news/uk-scotland-edinburgh-east-fife-22238457

Link: Baby rhino at Chester Zoo – breeding programme provides rhino to re-populate Kenya's Tsavo reserve

http://www.chesterzoo.org/must-sees/zoo-news/baby-rhino-embu

Explain to children that a generous benefactor has some money that they wish to donate to help an endangered or threatened animal survive and increase in numbers. They have asked the class to help them make the difficult decision about which animal to help. Children will need to make the case for an animal of their choice, providing evidence that they find from secondary sources of information to support their argument and presenting their findings in a creative way to the benefactor (or their representative) at the end of the lesson.

Challenges 1 and 2 are structured to provide appropriate support for children as they gather evidence and prepare arguments for the benefactor. Challenge 3 requires children to analyse information, use evidence and make judgments about the biggest threats to animals, from a given 'shortlist' of threats.

The challenges are presented on the Challenge slides to be displayed on the board, or printed out and placed in the centre of the table.

Challenge 1: Children collect evidence to argue a case, identifying responses to questions that might be useful in shaping a strong argument to make to the benefactor

Show Video 1 again. Ask them to make notes based on what they hear, which will help them to argue the case for supporting one or other of these magnificent creatures.

Provide them with a Planning grid (Resource sheet 1) on which to make notes in answer to questions about one of the animals. When they have added all they can to the grid, ask them whether they still need more evidence. Where might they find out more information? What do they need to know?

Once the children have collected as much evidence as they can, ask them to decide which are the three strongest pieces of evidence that will need to be included in their presentation to the benefactor.

Challenge 2: Children collect evidence to argue a case, using the prompts on the Challenge handout to help them create a strong argument to make to the benefactor

Provide the children with a list of endangered animals: peregrine falcon, bumblebee, salamander, osprey and koala bear. Ask them to choose one animal to find out about. Give them the Challenge handout (Resource sheet 2) and remind them to think carefully about the prompts. These will help them to target their research so that they make the best possible argument for their chosen animal.

Once the children have collected as much evidence as they can, ask them to decide the three strongest pieces of evidence that will need to be included in their presentation to the benefactor.

Challenge 3: Children consider a variety of threats to animal survival and create a 'threat map' that summarises what they have learned about these possible threats and their severity

Ask: *What are the biggest threats to animals?*

Give the children a list of six threats to animals (Resource sheet 3) and ask them to create a 'threat map' detailing what they know about how these six threats (and any others that they know about) impact on animals and their life cycles. If the children are not very experienced at mapping their ideas, provide them with prompts to help direct their thinking. For example: Loss of habitat > What sort of habitat? > Which animals might be affected by loss of habitat? > What problems would loss of habitat cause?

REFLECT AND REVIEW:

Revisit the Decision alley (Slideshow 1). Ask children to look back at what they thought then and see whether they would like to add to, or change, a response they made earlier. Ensure that children can justify any changes of viewpoint by explaining the evidence that they have since seen and why that evidence has made them change their mind.

Provide children with the opportunity to make their cases to the benefactor or their representative. It may be useful to shortlist the presentations beforehand, so that not all make the final 'event'.

EVIDENCE OF LEARNING:

Listen to children as they carry out research and share what they have discovered about the ways in which humans are using science to help endangered animals complete their life cycles. Can children identify reasons why animals might become threatened or endangered? Can they talk about one particular animal, telling its story in some detail? Can they describe ways in which humans are using science to help threatened or endangered animals, and recognise the positive and negative attitudes of some people towards, for example, breeding animals in zoos to protect species? Can children present evidence effectively to support an argument?

CIRCLE OF LIFE

ENRICHMENT LESSON 1: WHY DO ANIMALS MAKE INCREDIBLE JOURNEYS AS PART OF THEIR LIFE CYCLES?

Key vocabulary:

life cycle, humpback whale, blue whale, swift, osprey, wildebeest, caribou, monarch butterfly, migrate, migration, navigate, genetic

Resources:

Secondary sources for research, including quality non-fiction books, web-based resources, CDs, smartphone and tablet Apps, identification guides and leaflets

Key information:

Amphibians travel the shortest distances. They tend to migrate within their local area, often negotiating busy roads en masse so that they can return to the water courses from where they originated. Other types of animals migrate considerable distances during their lives. Not all migration, however, is linked to the animal's life cycle. For some, for example, salmon or turtles, it is about returning to specific places to breed. For others, the migration is driven by the need for food to sustain both adults and young animals.

LESSON SUMMARY:

In this lesson children find out about the incredible journeys that are undertaken by different types of animals during their life cycles. By the end of the lesson they will be able to describe in detail some of the journeys they find out about, giving reasons why the animals travel such huge distances.

National curriculum links:

Explain the differences in the life cycles of a mammal, an amphibian, an insect and a bird

Learning intention:

To explore some of the incredible migration journeys undertaken by animals as part of their life cycle

Scientific enquiry type:

Finding things out using a wide range of secondary sources of information

Working scientifically links:

Identifying scientific evidence that has been used to support or refute ideas or arguments

Success criteria:

- I can describe an incredible journey that a particular animal makes as part of its life cycle.
- I can give some reasons for that animal's migration and use evidence that I have selected to support my argument.
- I can compare the distances travelled by animals of very different sizes.
- I can suggest how an animal navigates during its incredible journey.

EXPLORE:

Show children Journey of the humpback whale (Video 1), which is based on storyline extracts from *The Journey of Humpback Whales* by Andy Belcher. Then play the True or false game (Interactive 1), encouraging children to think about the information they have seen in the video, as well as any prior knowledge about the life cycles of mammals that they may recall from earlier lessons.

ENQUIRE:

Explain to children that they are going to work together to answer questions about some incredible journeys.

They investigate the migration routes and distances travelled by a variety of animals, for example, swift, osprey, wildebeest, monarch butterfly, blue whale and salamander. The four types of animals that they have already studied are represented, but the distances that these animals travel as they migrate vary hugely.

Ask: *What do you think makes a journey 'incredible'? Is it the distance travelled, the time it takes, the size of the animal, or the difficulty or danger of the journey? Why do you think some animals travel huge distances as part of their life cycle? What difficulties and dangers might they face along the way? How do you think that migrating animals know where to go and how to get back?*

In Challenges 1 and 2 children are provided with different scaffolds to structure their research (using secondary sources of information) into the incredible journeys made by different animals. In Challenge 3 children independently identify a question to answer and evidence that they will need to collect in order to answer that question.

The challenges are presented on the Challenge slides to be displayed on the board, or printed out and placed in the centre of the table.

Challenge 1: Children create Key facts cards about animals that undertake 'incredible journeys', using information selected from secondary sources

Provide the children with the Key facts cards (Resource sheet 1) for two of the animals to refer to as examples. Ask them to use secondary sources of information, including quality non-fiction books, to help them identify and select relevant information to use as they create Key fact cards for other animals, for example, swift, osprey, wildebeest, blue whale or salamander.

Ask: *Which of the animals you have researched travels the furthest? Which is the smallest of these animals? Do smaller animals seem to travel further or less far than bigger animals? Does the size of the animal appear to make a difference?*

Challenge 2: Children use secondary sources of information to identify key data to add to a partially complete data set and then plot the information about the routes of each 'incredible journey' on an outline world map, using the knowledge they have gained to answer questions

Provide the children with a partially complete set of data about the incredible journeys of the animals they are investigating (Incredible journeys data table; Resource sheet 2). Explain that they need to use secondary sources of information to identify and select data that is missing from the grid. After they have completed the data collection, the children should plot the route of each incredible journey on an outline map of the world (Incredible journeys world map; Resource sheet 3).

Ask: *At what stage in their life cycle do the animals migrate? What are the reasons for the different journeys? Are there any patterns in your data?*

For example, the children might notice that females travel to a particular place to mate, have their babies and overwinter with their young, and then return to a place where there is more food and warmer weather.

Ask: *Is the reason for migrating always connected to the animal's life cycle or do other variables cause animals to migrate? What evidence shows you that migration is not only related to the animal's breeding cycle?*

Challenge 3: Children use a wide variety of secondary resources to identify relevant evidence that enables them to answer a question they identify for themselves

Challenge the children to find the answer to the following question: How do animals that make incredible journeys know where to go and how to get back?

The children should use a wide variety of secondary sources of information to help them investigate their ideas. If necessary, provide some prompts from the Clue cards (Resource sheet 4) to stimulate their thinking.

Ask the children to decide how best they can record the information that they research, relating what they discover to their knowledge and understanding of animals and their life cycles.

Key information:

Scientists believe that how animals navigate a migration route depends to a large extent on the type of animal, where it is going and why. Most animals are thought to use a combination of navigation methods and genetics is believed to play an important part. Research scientists look at data collected from years of observations of animal movements to help them understand behaviour patterns.

REFLECT AND REVIEW:

How do animals that make incredible journeys know where to go and how to get back? Ask children who completed Challenge 3 to feed back the results of their research, sharing the key points with the class.

Work with children to compare two animals that migrate, for example, a caribou and a swallow. Use the Venn diagram (Resource sheet 5) as the basis for comparing the two animals' migration journeys.

Ask: *How are their migration journeys similar? How are they different? Do they migrate for the same reasons? What reasons? Does the size of the animal make a difference in any way? Are the reasons for migration firmly related to the life cycle of the animal or are they sometimes related to other things?*

Compare other migratory animals and see if the answers are the same or different.

EVIDENCE OF LEARNING:

Listen to children as they carry out research and share what they have discovered about the migratory journeys animals make as part of their life cycles. Can they describe a migratory journey that a particular animal makes as part of their life cycle, providing details about the route taken, the distances travelled, etc.? Can they explain the reasons for that animal's migration and that of other animals, recognising that different animals have different reasons for making their journeys? Can children analyse the information that they have collected, making comparisons between the incredible journeys that the animals make? Can children, and particularly those who completed Challenge 3, suggest how an animal might navigate a route during its incredible journey?

REPRODUCTION IN PLANTS AND ANIMALS

INTRODUCTION

In this module children learn about reproduction in some types of plants and animals, including humans. This module should be taught after Module 1, The Circle of Life, as it builds on the learning about different types of animals and their life cycles begun during that unit. It also links closely with OCW, where children have opportunities to investigate and enquire practically into many aspects of the learning about reproduction in plants and animals that is the focus for this module.

As they learn about plant reproduction children will extend their knowledge from Year 3 of the function of the different parts of flowering plants. They will also learn that plants can reproduce in other ways, through asexual reproduction. As they learn about reproduction in animals children will find out more about specific mammals, birds, insects and amphibians and how they reproduce. There are three lessons focusing on humans, one of which is about the complete human life cycle and two of which focus on puberty. These lessons can be taught to mixed or single gender groups, but all children should learn about changes in boys and girls.

When working scientifically, children carry out first-hand observation of flowering and other plants, and also use secondary sources of information. They group and classify living things according to similarities in reproduction processes. They also report and present findings from their enquiries in a variety of ways, including posters, fact cards and guides.

Key vocabulary:

reproduction, reproduce, flower, organ, carpel, stamen, pollen, seeds, seed head, berry, fruit, pollinator, pollination, fertilisation, reproduction, reproduce, propagate, stem, leaf and root cuttings, runners, tubers, bulbs, rhizomes, gender, male, female, sex, sexual, asexual, metamorphosis, mate, sperm, pregnant, give birth, young, pup, calf, foal, chick, hatch, fledge, fledgling

FACT FILE:

Sexual reproduction in flowering plants

The reproductive organ of flowering plants is the flower.

The broad term 'flower' can be used to describe both **simple** and **compound** flowers. A simple flower has petals and contains a single set of reproductive parts at the centre, such as a buttercup or lily. Compound flowers appear to be single flowers, but the flower itself is actually made up of numerous small flowers arranged within a flower head. Daisies, dandelions and sunflowers are good examples of this. Most flowering plants have flowers with both male and female parts – 'perfect flowers' such as apple, tulip, daisy, dandelion and rose.

Some plants have separate male flowers and female flowers on the same plant, such as corn, courgette, marrow, squash and cucumber.

A smaller number of plants have male flowers and female flowers on separate plants, such as willow, maple and holly.

Children should learn that all plants do not produce 'perfect flowers' with both male and female organs, but that there are some plants with different sex flowers on the same or separate plants.

The female part of a flower consists of the **carpels**, which is where the seeds are formed. It has three parts: the **stigma**, the **style**, and the **ovary**. The male parts of the flower are the **stamens**, which produce **pollen**. Each stamen has two parts: an **anther** and a **filament**. The **anther** contains the pollen and the **filament** supports the anther.

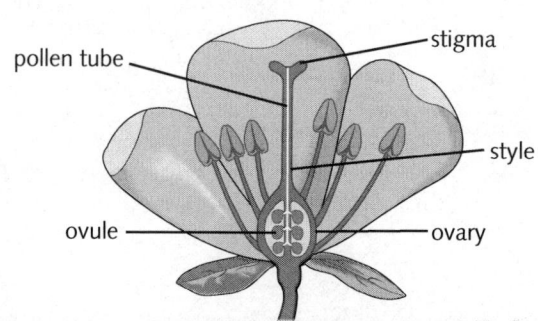

When the flower is **pollinated**, a pollen grain sticks to the stigma. It then travels through a narrow tube which grows down through the style to the **ovary**. In the ovary, the pollen joins with the **ovules**. This fusion of the male and female cells is called **fertilisation** and the fused cells divide to develop into **seeds**. After fertilisation, the ovary usually swells and becomes the **fruit**.

Asexual reproduction in plants

Many plants can also reproduce without forming seeds. This is called asexual or vegetative reproduction, which results in new plants that are genetically identical to the parent.

Plants may reproduce themselves naturally:

• Below ground – rhizomes, tubers, bulbs and corms. These are underground growths on the root or stem of a plant and contain stores of food to provide for the growing young plant.

• Above ground – the parent plant produces runners and new plants sprout along its length.

Reproduction in animals

One of the major characteristics of the major groups of animals is the means by which they reproduce. Although there are examples of animals that can reproduce asexually, this is not common. Sexual reproduction is therefore the norm in animals.

All **amphibians** reproduce by sexual reproduction. The female's eggs are fertilised outside her body. The female lays many jelly-covered eggs in water. The male immediately fertilises the eggs by spraying a cloud of sperm over them. Sperm enter into the eggs and fuse with the nucleus in the egg. The egg cells divide and develop until the offspring in the form of tadpoles are ready to hatch.

All **insects** reproduce sexually, but females of some species can also reproduce without the involvement of a male. This is called parthenogenesis and is an example of asexual reproduction rather than sexual – stick insects, often seen in classrooms, are a familiar species that can reproduce in this way.

All **birds** reproduce by sexual reproduction – sperm from the male fertilises the female's eggs inside her body. All **mammals** reproduce by sexual reproduction – sperm from the male fertilises the female's egg inside her body.

Common misconceptions:

• Children may think that humans are not animals and substitute 'animal' for types of mammal.

• Children tend only to recognise common mammals as animals and do not include birds, insects, fish and amphibians.

• Children may not appreciate that different types of animals have different life cycles; for example, they may think that all young animals start life as miniatures of their adult parents.

• Children may not recognise that reproduction is a characteristic of living things.

• Some children think that plants do not reproduce sexually at all.

• Children may think that bees and other insects visit flowers to pollinate them. They visit flowers to collect nectar; their role in pollination is accidental as far as the insect is concerned.

• Children may think that bees fertilise flowers; they pollinate them. Fertilisation happens when male and female genetic material fuses.

• Young children may not link the idea of mating and fertilisation to the birth of animals.

Big Cat book links

Life Cycles Sally Morgan 978-0-00-733640-1 Band 16 Sapphire	Explore the fascinating life cycle of the Salmon in this highly photographic book.

REPRODUCTION IN PLANTS AND ANIMALS

Key vocabulary:

reproduction, reproduce, flower, organ, carpel, stamen, anther, filament, pollen, seeds, seed head, berry, fruit, pollinator, pollination, fertilisation, sexual, asexual

Resources:

Enough flowers for at least one between two children. Ensure that the flowers are large enough to have identifiable male and female organs, such as alstroemeria or daffodils (you can also use lilies, unless children dissected these in Year 3, Module 1, Lesson 8). Magnifiers, digital microscopes, iPads, digital cameras

Health and safety:

Many plants produce wind-borne pollen which can cause hay fever and trigger an asthma attack.

Key information:

The **carpel** is the female part of the flower, where the seeds are made. The carpel has three parts: the **stigma**, the **style** and the **ovary**.

When the flower is **pollinated**, a pollen grain sticks to the stigma. It then travels down the style to the **ovary**. In the ovary, the pollen joins with the **ovules**, and the ovules become seeds. This is called **fertilisation**. After fertilisation, the ovary turns into the **fruit**.

The **stamens** are the male parts of the flower. Their job is to make **pollen**. Each stamen has two parts: an **anther** and a **filament**. The **anther** contains the pollen and the **filament** holds up the anther.

LESSON 1: HOW DO FLOWERING PLANTS REPRODUCE?

LESSON SUMMARY:

In this lesson children revise and build on work from Year 3, Module 1, How Does Your Garden Grow?, about the part that flowers play in the life cycle of flowering plants.

By the end of the lesson they will have a more detailed knowledge of the role of the flower, its parts and their function, and of the processes of pollination and fertilisation. They will have communicated their understanding of the process of sexual reproduction in flowering plants in a storyboard or other form.

This lesson links to and should be taught before OCW, Lesson 1.

National curriculum links:

Describe the life process of reproduction in some plants and animals

Learning intention:

To describe the process of sexual reproduction in many flowering plants, naming parts of the flower and explaining their importance within the process

Scientific enquiry type:

Grouping and classifying

Working scientifically links:

Reporting and presenting findings from enquiries, including conclusions, causal relationships and explanations of and degree of trust in results, in oral and written forms such as displays and other presentations

Success criteria:

- I can describe the life process of sexual reproduction in flowering plants, including pollination and fertilisation.
- I can explain the role of different parts of the flower in sexual reproduction.
- I can present the process of sexual reproduction in flowering plants in a storyboard or other sequence.

EXPLORE:

Ask: *What do we know already about how flowering plants reproduce?*

Play True/False/Not always with children using Resource sheet 1. Read the cards in turn, display them and write down children's responses. This will provide a useful opportunity for you to assess children's existing knowledge and understanding, along with any misconceptions they may have. This card sort is used here and again at the end of the lesson. Statements included at this point revise children's learning about reproduction in flowering plants from Year 3. Additional cards should be added for the Reflect and Review part of the lesson, to help children recognise how their understanding has developed.

ENQUIRE:

Children should learn that all flowering plants do not produce 'perfect flowers' complete with male and female organs, but that there are some flowering plants with different sex flowers on the same or separate plants.

Reinforce the importance of the flower. In this lesson the focus is on 'perfect flowers', those that contain both male and female parts. In the next lesson children will look closely at flowering plants that have separate male and female flowers on the same or different plants.

Show children Parts of the flower (Animation 1). Check their understanding of the processes of pollination and fertilisation that are shown.

Ask: *Where is pollen produced? How is it transferred onto the female part of the plant (the carpel)? What happens then?*

Reinforce vocabulary and ensure that children have a clear understanding of terms such as 'pollination' and 'fertilisation'.

Explain that they are going to use a real flower to communicate the story of reproduction in flowering 'perfect plants'. All children will need a flower, which they can draw/photograph as they disassemble it, focusing on each stage of the process in turn. As children dissect each part, ask them to consider these key questions each time: What is this part? What is its function?

Key information:

Most flowering plants have flowers with both male and female parts – 'perfect flowers' such as apple, tulip, daisy, dandelion and rose.

Some plants have separate male flowers and female flowers on the same plant, such as corn, courgette, marrow, squash and cucumber.

A smaller number of plants have male flowers and female flowers on separate plants, such as willow, maple and holly.

Challenges are differentiated according to the level of support children are given, as they communicate the stages of the process of sexual reproduction in flowering plants. In Challenges 1 and 2 children are provided with an appropriate storyboard grid to help them order the steps of the pollination and fertilisation processes, while in Challenge 3 children are encouraged to use a method of communication of their own choice but are prompted to ensure that sufficient detail is included.

The challenges are presented on the Challenge slides to be displayed on the board, or printed out and placed in the centre of the table.

Challenge 1: Children dissect and draw the whole flower and its parts, and describe the steps of the processes of pollination and fertilisation

Provide the children with an enlarged version of the Storyboard grid 1 (Resource sheet 2) with prompts to help them as they sequence the steps of the processes of pollination and fertilisation. The children should use drawings of their flower parts to illustrate their explanations of each step of both processes. Encourage them to watch the animation sequence again to refresh their memories and ensure that they are clear about the later stages of the process.

Challenge 2: Children dissect and draw or photograph the flower and the parts of the flower, describing the steps of the processes of pollination and fertilisation

Provide the children with an enlarged version of the Storyboard grid 2 (Resource sheet 3) with limited prompts indicating key steps of the sequence. They should use this to help them structure the process of pollination and fertilisation in flowering plants with 'perfect' flowers. The children should use a combination of photographs of their flower as it is disassembled and drawings to illustrate their description and explanation of each step of the process.

Challenge 3: Children choose the best way to communicate their understanding of the process of pollination and fertilisation.

The children have a flower each and may use ICT and other resources to communicate the sequence of events involved in pollination and fertilisation. Encourage them to be creative, perhaps using Photo Story or PowerPoint, and to include photographs or sound clips, as well as text. They will need to include the names of all parts of the flower and their function, detail of the male and female parts, and the processes of pollination and fertilisation.

REFLECT AND REVIEW:

Ask: *How will we know that we have done a good job in communicating our understanding of sexual reproduction in some flowering plants? What should all our storyboards or presentations include?*

Ensure that children mention the names of the parts of the flower and their function, the process of pollination, the process of fertilisation and the making of seeds.

Repeat the True/False/Not always game children played earlier in the lesson. Can children now place any statements that they were not sure about earlier and do they want to move any? What evidence do they have to justify these moves? Add these further statements, which will challenge their understanding further and stimulate curiosity about what they are going to learn about in the next lessons.

Only some flowers can be pollinated: True – male flowers produce pollen, but they cannot be pollinated

Plants reproduce by making seeds that grows into a new plant : Not always

All flowers produce pollen: False – not female flowers

All flowers have male and female parts: False – some flowers have only male or only female parts

Some plants can reproduce in ways other than making seeds: True – almost all produce seeds, but they can reproduce in other ways. Some do not produce seeds; for example, ferns and mosses produce spores. Many plants can also reproduce asexually, which does not involve seed production.

EVIDENCE OF LEARNING:

Listen to children's responses. Can they describe how flowering plants reproduce sexually? Do they know key vocabulary, such as names of flower parts and processes involved in reproduction? Do they know what a 'perfect flower' is? Are they able to use the key vocabulary effectively to communicate their understanding? Can they explain the role of different parts of the flower in sexual reproduction? Can they evaluate how well they have communicated their understanding of sexual reproduction in flowering plants?

REPRODUCTION IN PLANTS AND ANIMALS

Key vocabulary:

reproduction, reproduce, flower, organ, carpel, stamen, pollen, pollinator, pollination, fertilisation

Resources:

A variety of flowers (different from those observed in Lesson 1, including some single sex flowers), such as courgette, marrow, holly. If none are available, use images of single sex flowers, magnifiers, digital microscopes, digital cameras, modelling clay, clay, junk modelling resources

Health and safety:

Grasses, catkins and many other plants produce wind-borne pollen which can cause hay fever and trigger an asthma attack. Avoid working with these plants inside the classroom if there is a risk of the release of large amounts of pollen.

Key information:

Most flowering plants have flowers with both male and female parts – 'perfect flowers' such as tulip, daisy and rose.

Some plants have separate male and female flowers on the same plant, such as corn and courgette.

A smaller number of plants have male and female flowers on separate plants, such as willow, maple and holly.

Children should learn that all flowering plants do not produce 'perfect flowers' complete with male and female organs, but that there are some flowering plants with different sex flowers on the same or separate plants.

Ⓒ LESSON 2: ARE ALL FLOWERS ON ALL PLANTS THE SAME?

LESSON SUMMARY:

In this lesson children further develop their understanding of the role of flowers in the reproductive cycle of plants. By the end of the lesson they will have learned that not all plants have 'perfect flowers', containing both male and female parts, but that some have male and female flowers that differ in shape, size and structure, on the same or different plants of the same species. This lesson links to and should be taught before OCW, Lesson 2, where children will explore plants with flowers of different types, observing the flowers in detail as they change across the year.

National curriculum links:

Describe the life process of reproduction in some plants

Learning intention:

To recognise that flowers are not all the same and identify how they are different

Scientific enquiry type:

Grouping and classifying

Working scientifically links:

Identifying scientific evidence that has been used to support or refute ideas or arguments

Success criteria:

- I can identify, name and describe in detail the function of the reproductive parts of a flower and the processes involved in plant reproduction.
- I can describe how the flowers of some plants are different, including only male or female parts.
- I can explain how the process of reproduction differs in plants that have single sex flowers.

EXPLORE:

Display the four statements on the first part of Interactive 1 and ask which are true and which false. Go on to the second part of Interactive 1 and click on the boxes to reveal explanations of the answers.

Remind children that in the last lesson they learned about reproduction in most flowering plants. Ask them to talk to their partners and come up with a description of a flower and the process of reproduction. Ensure that they mention and name the male and female parts of the flower, such as stamen and carpel, and describe their function. Remind them of the term 'perfect flower' to describe a flower that has both male and female organs.

Explain that today they are going to learn about some other types of flowering plants.

Show them the Male and female flowers (Video 1), which gives a number of examples of other arrangements of flowers on plants: separate male and female flowers on the same plant, male and female flowers on separate plants of the same type.

Ask: *What did you notice about male and female flowers from the same plant? How are they different from 'perfect flowers' – those that have male and female parts?*

ENQUIRE:

Explain that they are going to work with a partner to create a model to help explain the difference between male and female flowers from the same species of plant.

Challenges are differentiated according to the level of support the children are given as they plan a design for a model of male and female flowers. In Challenge 1 the children are provided with a planning grid to help them identify and plan the parts they need to include in both male and female flowers. In Challenges 2 and 3 the children are expected to design model flowers that reflect the particular arrangement of flowers in different types of plant. All the children evaluate the quality of their designs against success criteria they develop during the Reflect and Review part of the lesson.

The challenges are presented on the Challenge slides to be displayed on the board, or printed out and placed in the centre of the table.

Challenge 1: Children use a planning grid to help them design a model to explain the differences between male and female flowers

Provide the children with the Planning grid (Resource sheet 1) on which to plan and sketch their design. They should specify on their design the names of the flower parts they need to include and their arrangement in each flower. The children may plan and produce a mock-up of their model, explaining what it shows during the lesson, and complete the model itself during a longer session.

Challenge 2: Children sketch designs for a model of separate male and female flowers that are located on the same plant

They should specify on their design the names of the flower parts they need to include and their arrangement in the flower. The children may plan and produce a mock-up of their model, explaining what it shows during the lesson, and complete the model during a longer session.

Challenge 3: Children sketch designs for a model of a male and a female flower, where the flowers of different sexes are on separate plants

They should specify on their design the names of the flower parts that they need to include and their arrangement in the flower. They should also think about how the female flower will be pollinated and the seeds fertilised, and how they can model this effectively. The children may plan and produce a mock-up of their model, explaining what it shows during the lesson, and complete their model during a longer session.

REFLECT AND REVIEW:

Ask: *How will we know if our models help to explain the process of reproduction in plants with male and female flowers?*

Support children to come up with their own success criteria for a good model. Ensure that children mention the names of the parts of the flower and their function, the position of those organs in single sex flowers, the different arrangement of flowers on plants which do not have 'perfect flowers', the processes of pollination and fertilisation, and the making of seeds.

Ask a selection of children to share their models (or the designs for their models), describing the structures and processes they have chosen to include. Ask the others to listen carefully, evaluate using their success criteria, and give feedback to their classmates.

EVIDENCE OF LEARNING:

Listen to children as they produce their model design and during the Reflect and review part of the lesson. Do they have a clear understanding of the function of the reproductive parts of a flower and the processes involved in plant reproduction? Do they use key vocabulary accurately as they describe and explain the flowers they design? Can they explain how the process of reproduction differs in plants that have single sex flowers from that in perfect flowers and give examples?

REPRODUCTION IN PLANTS AND ANIMALS

Key vocabulary:

reproduce, propagate, stem, leaf and root cuttings, runners, tubers, bulbs and rhizomes, asexual, vegetative

Resources:

Examples of bulbs such as garlic, onions or shallots (some of which can be cut up), tubers, rhizomes, seed potatoes, plants in pots such as fuchsia, begonia, geranium, rosemary, mint, strawberry

Health and safety:

Wash hands after handling plant material. Some plants may have sap which will irritate a child's skin. Some bulbs, such as hyacinths, can also cause skin irritation.

Key information:

Reproducing other than by producing seeds saves a plant energy, although the process of seed production and dispersal ensures more genetic diversity within the population.

Plants may reproduce themselves:

Below ground – rhizomes, tubers, bulbs and corms. These are underground growths on the root or stem of a plant and contain stores of food to provide for the growing young plant.

Above ground – the parent plant produces runners and new plants sprout along its length. These plants are 'clones' of the parent.

From stem, root or leaf cuttings – when pieces of plant matter are placed in soil or water, new plants, complete with root systems, are produced.

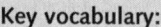

LESSON 3: DO ALL PLANTS REPRODUCE BY PRODUCING SEEDS?

LESSON SUMMARY:

In this lesson children learn about asexual reproduction, that is, the ways that plants can produce new plants from different parts of the parent plant, rather than by producing seeds. By the end of this lesson they will be able to describe the different methods in some detail and suggest the benefits to plants of asexual reproduction.

National curriculum links:

Describe the life process of reproduction in some plants

Learning intention:

To describe how plants can reproduce asexually, by creating new plants from different parts of the parent plant rather than by producing seeds

Scientific enquiry type:

Finding things out using a wide range of secondary sources of information

Working scientifically links:

Reporting and presenting findings from enquiries, including conclusions, causal relationships and explanations of and degree of trust in results, in oral and written forms such as displays and other presentations

Success criteria:

• I can describe ways that plants are able to reproduce other than by producing seeds.

• I can suggest why it is useful for plants to reproduce asexually.

• I can produce a step-by-step 'How to' guide for a propagation method of my choice.

EXPLORE:

Begin with a number of prompt questions.

Ask: *Why is it useful for plants to reproduce without producing seeds? What needs to take place for seeds to be produced? Can you think why this could sometimes be a problem?*

Show children Propagating plants (Video 1), which describes how plants can reproduce themselves, other than by producing seeds.

Ask: *What different methods of plant reproduction did we hear about? Has anybody propagated plants by taking cuttings, planted bulbs or grown potatoes themselves?*

Encourage children to talk about their experiences of gardening at home or with a relative, or in other classes at school or gardening club.

ENQUIRE:

Provide a selection of real plants and bulbs for children to look at, or alternatively use the images from Resource sheet 1. Encourage children to think about sorting the plants into groups, based on how they reproduce, other than by producing seeds. They should use the headings (cuttings – artificial; above ground – natural; below ground – natural) and information from the resource to organise their sort. For example: fuchsias – cuttings; strawberries – above ground; potato – below ground. Once this classification is complete, ask them to sort again, but this time in more detail, sub-dividing each group into sub-groups. The purpose of this is to ensure that children become more aware of the variety of ways that plants can reproduce and propagate themselves.

Explain that during the rest of this lesson they will be finding out about a particular method of asexual reproduction in more detail, producing a step-by-step 'How to' guide to help gardeners propagate plants using the instructions that they produce. Challenges are differentiated according to the level of structure and support provided to children as they identify and select key information to include in a 'How to' guide.

The challenges are presented on the Challenge slides to be displayed on the board, or printed out and placed in the centre of the table.

Key information:

Propagation methods are frequently used to reproduce plants artificially, such as by taking stem, root or leaf cuttings.

Many plants produce 'runners' that spread naturally, with roots being produced at points where a stem bends downward and comes into contact with the earth, such as blackberries. Some plants also create new individual plants underground in this way, producing lateral roots that bud and form new individual plants, such as buttercups.

Challenge 1: Children use a writing frame to identify and plan six steps that explain how to grow potatoes in a container

Provide the children with secondary sources of information about how to grow potatoes from tubers (in pots or containers) (thompson-morgan.com/how-to-grow-potatoes-in-bags)

Ask them to work with a partner, using the writing frame (Resource sheet 2) to help them select key information to record, and limiting their instructions to six steps only. They can use diagrams or drawings to help explain each step if they wish.

Ask: *How will you decide which information is the most important to include?*

Ensure that they are aware that they need to read all the information first, before highlighting the most crucial things to include in their six simple steps.

Challenge 2: Children select information to include in a guide about propagating plants by taking stem cuttings

Provide the children with a variety of secondary sources of information, including web-based material, leaflets and brochures from garden centres and plant suppliers, about how to take stem cuttings. Ask them to produce a step-by-step 'How to' guide to the process, using the planner (Resource sheet 3) as a possible structure. They may choose to present their instructions in a different way if they wish, but should be limited to one side of A4 paper for their guide.

Ask: *How will you decide which information is the most important to include?*

Challenge 3: Children research information and make notes to produce a 'How to' guide

Explain to the children that sometimes parts of plants in the natural world – perhaps a branch, stem or leaf – touch the soil. This encourages that part of the plant to grow its own roots and become a separate plant.

Ask: *How might a commercial grower of plants use this knowledge to create replica plants?*

Allow the children time to think about this and prompt them, if necessary, to think about layering of runners to produce new strawberry plants and the taking of leaf cuttings, such as begonia and violets. The children should research key information, making notes that they can then use to produce a step-by-step 'How to' guide for others to use.

REFLECT AND REVIEW:

Review with children what they have learned during the lesson so far and ask them to share propagation methods they have found out about with others. Ensure that they are clear that the methods they have found out about during this lesson are asexual forms of plant reproduction that can sometimes take place naturally, but are more often used artificially – especially by crop producers – to produce large numbers of plants quickly.

Explain that they are now going to look at another way some plants reproduce themselves asexually. You are going to look at what happens inside a bulb.

Have ready a visualiser and a large bulb that you have sliced in half, and a selection of bulbs that children can hand around and look at more closely (such as garlic or shallots). Place the halved bulb on the visualiser, so that the detail of the layers can be seen. Explain that bulbs contain everything that the plant needs to allow it to grow. For this reason bulbs are very successful where the soil is poor, as they need no additional nutrients.

Ask: *How should we plant bulbs? Does anybody know already?*

Challenge children to find out for homework.

Key information:

A bulb is a rounded underground storage organ, consisting of a short stem surrounded by fleshy scale leaves or leaf bases, which lies dormant over winter.

EVIDENCE OF LEARNING:

Listen to children's responses throughout the lesson. Review their 'How-to' guides. Can they describe in detail at least one way that plants are able to reproduce other than by producing seeds? Can they explain the differences between sexual and asexual reproduction? Are they able to suggest why it might be useful for plants to reproduce in a variety of different ways? Did they produce a step-by-step 'How to' guide that describes a propagation method clearly, so that another child would be able to follow the instructions?

REPRODUCTION IN PLANTS AND ANIMALS

LESSON 4: HOW DO AMPHIBIANS AND INSECTS REPRODUCE?

LESSON SUMMARY:

In this lesson children find out in more detail about how amphibians and insects reproduce. By the end of this lesson they will be able to compare the process of reproduction in amphibians and insects, identifying and describing similarities and differences between the two and recognising both as examples of sexual reproduction, with some exceptions. This lesson links with and should ideally be taught after Lesson 3 and 4 of Module 1, The Circle of Life. It also links closely with the following lesson in this module, Lesson 5.

Key vocabulary:

reproduce, reproduction, gender, male, female, sex, sexual, asexual, metamorphosis

Key information:

All **amphibians** reproduce by sexual reproduction. The female's eggs are fertilised outside her body. The female lays many jelly-covered eggs in water. The male immediately fertilises the eggs by spraying a cloud of sperm over them. The eggs then develop until they are ready to hatch. Amphibians spend the early stages of their lives in water, using gills to breathe, and the later stages on land as well as in water, developing lungs to help them breathe.

All **insects** reproduce sexually, but females of some species can also reproduce without the involvement of a male. This is called parthenogenesis and is an example of asexual reproduction rather than sexual – stick insects, often seen in classrooms, are a familiar species that can reproduce in this way. Most female insects mate with a male who fertilises their eggs. The female then lays large numbers of eggs – sometimes in water or sometimes on plants – where they develop without care from the parents.

National curriculum links:

Describe the life process of reproduction in some plants and animals

Learning intention:

To describe the life process of reproduction in amphibians and most insects and recognise this process as sexual reproduction

Scientific enquiry type:

Finding things out using a wide range of secondary sources of information

Working scientifically links:

Reporting and presenting findings from enquiries, including conclusions, causal relationships and explanations of and degree of trust in results, in oral and written forms such as displays and other presentations

Success criteria:

- I can describe how some amphibians and insects reproduce.
- I can identify metamorphosis as a stage in the life process of reproduction that is specific to these two types of animals.
- I can explain that amphibian and most insect reproduction is sexual reproduction, requiring two animals, one male and one female.

EXPLORE:

Group children in twos or threes. Give each group a collection of cards (Resource sheet 1) with key words they should recognise from lessons in Module 1, The Circle of Life, plus some blank cards. Ask them to organise the words on a large sheet of paper, joining the words with arrows and identifying linking phrases. If the children have not drawn a concept map before, or for a while, model a few linking examples to get them started, such as Eggs (are laid on) Plants, (they hatch into) Caterpillars, (change into) Butterflies (written on blank card).

After an appropriate time, ask the children to take it in turns to describe the links they have made. Comment on any different, more creative links that have been made, and note any misconceptions children might have.

Encourage children to think about how the life process of reproduction in amphibians and insects are similar and how they are different.

Ask: *What is different about reproduction in amphibians and insects? Where do they each lay their eggs? When does metamorphosis take place? Have you noticed anything similar about how they reproduce? Does reproduction always involve a male and a female?*

Explain that this is known as sexual reproduction, because a female and a male are involved in the process. Ensure that they are clear that it is similar to sexual reproduction in plants, which they learned about during earlier lessons in this module (Lessons 1 and 2).

ENQUIRE:

Explain that during this lesson they will be discovering more about sexual reproduction as a process in the life cycle of amphibians and insects. In Challenges 1 and 2 children select information to include in a 'Just a Minute' style presentation. In Challenge 1 children are provided with a list of questions to help focus their thinking while in Challenge 2 children select information to include themselves. In Challenge 3 children are challenged to try to make sense of the presence of baby stick insects, when only one female adult is present. They use their knowledge of sexual reproduction and research from secondary sources to help them explain this very unusual example of an insect that can reproduce asexually as well as sexually.

The challenges are presented on the Challenge slides to be displayed on the board, or printed out and placed in the centre of the table.

Challenge 1: Children answer a series of questions as they identify information to include in a minute's presentation about the reproduction of an amphibian or insect

Provide the children with a list of questions (Resource sheet 2) to prompt them as they think about sexual reproduction in either an amphibian or an insect of their choice.

Explain that their challenge is to find out enough information to fill 'Just a Minute' of talk time. Their subject is reproduction in an animal of their choice, either an amphibian or an insect.

They work with a partner to come up with answers to the list of questions, note down what they know already and check anything they are not sure about. To help them, they might look back at their learning from the earlier module, The Circle of Life, where children will have learned about several amphibian and insect life cycles. You may need to prompt their choice, ensuring that children across the class prepare information on a variety of different amphibians and insects.

Ask them to rehearse their minute before trying it out on other children. Classmates can question them for further information or challenge them if anything they say is wrong. – AP

Challenge 2: Children identify key information to include as they talk for 'Just a Minute' about reproduction in an amphibian or an insect they choose

They will need to work with a partner, think about what they know already, before checking anything they are not sure about. They should also include information about the animal's habitat. They should then rehearse their minute before trying it out on other children. Classmates can question them for further information or challenge them if anything they say is wrong. – KH

Challenge 3: Children use their knowledge of sexual and asexual reproduction to help them develop an explanation for an unusual event

Tell the children that Jamie, a boy from a local school, has discovered something of a mystery. His pet stick insect Betty lives alone in a special tank. Today, the tank is full of tiny baby stick insects – as well as Betty! What is the explanation for this? Where have the baby insects come from? What gender are they? Explain to the children that their challenge is to provide Jamie with the answers he needs. How do stick insects reproduce? How are they different from most other insects? The children will need to use secondary sources of information to help them develop an explanation. – DB

Key information:

Female stick insects can reproduce through parthenogenesis, meaning they can lay eggs that will hatch into clones of themselves. As a result, males are much less common than females in most populations. Males are only produced when a female chooses to mate, resulting in eggs that have a 50/50 chance of being male.

REFLECT AND REVIEW:

Begin by asking children who completed Challenges 1 or 2 to talk for 'Just a Minute' about their chosen animal. Ensure that the class hears a variety of minutes, covering amphibians as well as insects.

Ask: *How is sexual reproduction in amphibians and insects similar? How is it different?*

Ensure that children identify that (sexual) reproduction in both animals involves a male and a female. They may mention details in identifying differences, such as the location where eggs are laid and the different forms of metamorphosis that the animals go through.

Use the Compare and contrast grid (Resource sheet 3) to compare sexual and asexual reproduction in insects, specifically a butterfly and a stick insect. Encourage children who completed Challenge 3 to share what they have learned about the different ways that a stick insect can reproduce.

Focus on the similarities and differences between the two animals and encourage children to think about how they will find further information, if they do not know all the details required.

EVIDENCE OF LEARNING:

Listen to children's responses throughout the lesson. Are they able to describe in detail how some amphibians and insects reproduce? Can they identify 'metamorphosis' as a stage in the life process of reproduction that is specific to both but can happen differently, depending on the animal involved? Can they explain that reproduction in amphibians and most insects is sexual reproduction, because it involves one male and one female? Can children describe (in simple terms) asexual reproduction that often occurs in a stick insect?

REPRODUCTION IN PLANTS AND ANIMALS

LESSON 5: HOW DO MAMMALS AND BIRDS REPRODUCE?

LESSON SUMMARY:

In this lesson children find out more about how mammals and birds reproduce. By the end of this lesson they will be able to compare the process of reproduction in mammals and birds, identifying and describing similarities and differences between the two and naming both as examples of sexual reproduction.

This lesson links with Lessons 2 and 5 from Module 1, The Circle of Life.

National curriculum links:

Describe the life process of reproduction in some plants and animals

Learning intention:

To describe the life process of reproduction in mammals and birds and recognise this process as sexual reproduction

Scientific enquiry type:

Grouping and classifying

Working scientifically links:

Reporting and presenting findings from enquiries, including conclusions, causal relationships and explanations of and degree of trust in results, in oral and written forms such as displays and other presentations

Success criteria:

- I can describe how some mammals and birds reproduce.
- I can identify similarities and differences between the life process of reproduction in these two types of animals.
- I can explain that mammal and bird reproduction is a type of sexual reproduction, requiring two animals – one male and one female.

EXPLORE:

Show children Courtship rituals (Video 1), a video that includes examples of male and female mammals, male and female birds, and their courtship rituals.

Ask: *What do you notice about male and female birds? Which one is bigger?*
   (This is often the female.)
   Which is the most brightly coloured?
   (The male often has colourful, exotic-looking plumage.)
   What do the males do to attract a mate?
   (They may prepare a nest site or perform a courtship dance of some kind.)
   How does a male mammal attract a mate? What is important for a male mammal to show? How do they demonstrate their fitness and strength?
   (Sometimes by displaying – like some birds – sometimes by fighting.)

Establish that children recognise that female mammals and birds are looking for the best possible mate so that they will produce healthy young that stand a good chance of surviving and growing into adults.

ENQUIRE:

In Challenges 1 and 2 children use Venn diagrams to organise information about reproduction in mammals and birds. In Challenge 1 children are provided with statements to sort, while in Challenge 2 they generate, then sort, their own statements. In Challenge 3 children are asked to work collaboratively to identify evidence that helps them compare, contrast and rate the parenting skills of male mammals or birds.

The challenges are presented on the Challenge slides to be displayed on the board, or printed out and placed in the centre of the table.

Challenge 1: Children use a Venn diagram to compare reproduction in mammals and birds

Provide the children with a blank Venn diagram (Resource sheet 1) and a set of statement cards (Resource sheet 2, which includes some blanks for their own ideas) to help them compare and contrast reproduction in mammals and birds. Ask them to think about the statements and work together in pairs to complete the Venn diagram, putting each statement into the correct position. Warn them that some of the statements will be simple to place, but others may be harder. What is important is that they can explain why they have chosen to put a statement in a particular place. Encourage children to look at books and other secondary sources of information to help them to decide.

Challenge 2: Children compare reproduction in a specific mammal and bird

Provide the children with a blank Venn diagram (Resource sheet 1) to use to compare and contrast reproduction in mammals and birds. Ask them to choose a specific mammal and bird, such as an elephant or kangaroo and an ostrich or kestrel, and come up with a series of statements (recording these on cards or Post-it notes) about each animal. They may also choose to compare a bird or mammal they have observed first-hand, such as farm animals or pets.

Ask: *Are there statements that are true for both?* These go in the middle section. *Are there statements that are either about the bird or the mammal?* These go in one of the other sections. As they are close to finishing their task ask: *Are the statements that you have put in the 'bird' section true of all birds? Are there any examples that do no fit?* For example, the children may think that all birds lay lots of eggs, but some birds only lay one, such as penguins.

Challenge 3: Children compare the parenting skills of male mammals and birds and present results to the rest of the class

Ask: *Do you think that male mammals are better parents than male birds? What would you need to do to find out?*

Ask the children to look for and collect evidence, prepare a case and present it to the rest of the class. To share the task they can work in a small group or in pairs, but must come up with a plan to find the evidence they need to answer the question and make a case as a team.

REFLECT AND REVIEW:

This activity brings together children learning about reproduction in amphibians, insects, birds and mammals.

Show children Interactive 1, which features a series of statements and the headings Agree, Disagree and It depends. The children need to read the statements in turn; they can talk to their partners before deciding what they think. They need to be able to explain their thinking, justify their ideas, and be prepared to listen to others if they disagree with them.

Where children disagree (or agree) with a statement, ask for evidence. For example, 'All animals reproduce. Their eggs develop into young that can look after themselves'. Children might disagree. Ask them to give an example justifying their response, such as a tadpole feeds itself, but chicks have to be fed by their parents. You might respond in turn by asking for exceptions.

Ask: *Does anybody think differently? What about hen chicks in an incubator? Do their parents look after them?*

EVIDENCE OF LEARNING:

Listen to children's responses throughout the lesson. Are they able to describe how mammals and birds reproduce, and talk in detail about specific examples? Can they identify broad similarities and differences between the life process of reproduction in mammals and birds? Can they describe how some mammals and birds choose a suitable mate? Can they explain that mammal and bird reproduction is a type of sexual reproduction, requiring two animals – one male and one female? Do they recognise that almost all animals reproduce sexually, with a few exceptions?

REPRODUCTION IN PLANTS AND ANIMALS

LESSON 6: HOW DOES THE HUMAN LIFE CYCLE COMPARE WITH THAT OF OTHER MAMMALS?

Key vocabulary:

life cycle, birth, growth, reproduction, ageing, death, baby, toddler, teenager, adult, adulthood, childhood, pregnancy, gestation, puberty, sexual, mammal

Resources:

Mini whiteboards and pens

LESSON SUMMARY:

In this lesson children identify the stages of the human life cycle, including puberty and pregnancy, and compare lengths of gestation for different mammals. By the end of this lesson they will have plotted a scatter graph and used it to find a correlation.

National curriculum links:

Describe the changes as humans develop to old age

Learning intention:

To recognise patterns in data about the life cycles of humans and other mammals

Scientific enquiry type:

Noticing patterns

Working scientifically links:

Recording data and results of increasing complexity using scientific diagrams and labels, classification keys, tables, scatter graphs and bar and line graphs

Success criteria:

- I can name and sequence the stages on a human life cycle diagram.
- I can compare the human life cycle with that of other mammals.
- I can present data as a scatter graph.
- I can use my graph to answer a question about life cycles.

EXPLORE:

Show the life cycle diagram (Interactive 1) to children. Provide them with mini whiteboards and pens. For each familiar stage in turn (birth, growth, reproduction, ageing, death) ask children to write its name on their whiteboard before revealing it on the diagram (these will be the five labels nearest the top of the screen). Click below the reproduction label to reveal the arrow between reproduction and birth.

Ask: *What kind of animal are humans? What is the name for the stage between reproduction and birth in a mammal?*

Reveal the label for this stage. Move on to the next screen to show the pictures and choose children to place them on the life cycle diagram. Draw attention to the adult male and female. The picture includes both because mammals reproduce sexually; a male and a female are needed. You may also want to draw attention to the fact that the elderly female is past the age at which she can reproduce. Move on to the next screen to the label for childhood (this is just after the birth label).

Ask: *What is the name for the stage between childhood and adulthood?*

Reveal the label. If children have not already learned about puberty in SRE lessons, explain to them that it is the stage during which humans become ready to reproduce.

Key information:

Confirm that other mammals go through similar stages but that they take different amounts of time. Young animals are born with varying abilities to walk, see and so on, and are dependent on the mother for different lengths of time. All animals reach sexual maturity as they enter adulthood; the term 'puberty' is used for this life cycle stage in humans.

ENQUIRE:

Remind children that in Lesson 5 of this module and in Lesson 2 of Module 1, The Circle of Life, they learned about reproduction and life cycle in mammals. Ask them to look back at their previous work and then think, pair, share to identify similarities and differences between the human life cycle and that of other mammals.

EXPLORE:

Explain that in this lesson they will be comparing gestation, a stage in the life cycle that they have not yet considered. Show Slideshow 1.

Ask: *Can you see any patterns in the information about length of gestation?*

Children should notice that larger animals have longer gestation periods.

Ask: *What would we need to do to be more certain about the pattern we have seen?*

Confirm that they need to analyse more data.

ENQUIRE:

Advise children that their challenge is to use the data you give them to answer the question: Is there a relationship between length of gestation and the size of a mammal? The challenges are differentiated by the amount of data and the presentation and interpretation skills needed.

The challenges are presented on the Challenge slides to be displayed on the board, or printed out and placed in the centre of the table.

Challenge 1: Children present data about length of gestation as a simple scatter graph and use it to answer the enquiry question

These children work in pairs or small groups. Provide them with the animal cards (Resource sheet 1) and an enlarged copy of the graph template (Resource sheet 2). Ask the children to stick each card in the appropriate section of the graph then use the graph to answer the question: What is the relationship between length of gestation and the size of a mammal?

Challenge 2: Children present data about length of gestation as a scatter graph and use it to answer the enquiry question

These children work individually. Provide them with the table of data about gestation and weight (Resource sheet 3) and the graph template (Resource sheet 2). Ask them to plot the data on the graph then use it to answer the question: What is the relationship between the length of gestation and the weight of a mammal?

Challenge 3: Children present data about length of gestation as a scatter graph and decide how good an answer it provides to the enquiry question

These children work individually. Provide them with the table of data about gestation and weight (Resource sheet 4) and the graph template resource sheet (Resource sheet 5). Ask them to plot the data on the graph then use it to answer the question: What is the relationship between the length of gestation and the weight of a mammal?

Ask: *Are there any exceptions to the pattern? Do you think the pattern is likely to be true for all, most or some mammals?*

Ask: *Where on the graph does this animal go? What length of gestation and size would an animal have if it was here on the graph? What pattern can you see? Do humans fit the pattern?*

REFLECT AND REVIEW:

Share one of the graphs from Challenge group 3. Discuss what the graph shows.

Ask: *Does it show the same pattern that you found? Are there any animals which do not fit the pattern?*

Confirm that for many animals there is a correlation between size and length of gestation but there are exceptions, and these include humans and other primates.

Ask: *Could we investigate this question for any other group of animals? Why not?*

Explain to children that in the next two lessons they will be learning about puberty. Set up a question box where they can put any questions they have about puberty or the human life cycle. Some of these will be answered in Lessons 7 and 8 and others may be discussed in PSHE lessons.

EVIDENCE OF LEARNING:

Listen to children as they help sequence the human life cycle pictures. Observe as they compare the data. Can they name the stages of a generic life cycle? Do they know the stages of pregnancy and puberty? Can they correctly place the pictures on the human life cycle and name the stage they represent? Do children know that humans are mammals and that pregnancy/gestation is only a feature of mammalian life cycles? Can they present data on a scatter graph? Can they identify the pattern on the scatter graph and answer the enquiry question? Can they identify data which does not fit the pattern and decide how confident they are about the pattern?

CROSS CURRICULAR OPPORTUNITIES:

This lesson links with the school's PSHE scheme of work. It also links to Mathematics with graphs and data presentation.

REPRODUCTION IN PLANTS AND ANIMALS

LESSON 7: HOW DO GIRLS BECOME WOMEN?

LESSON SUMMARY:

This is the first part of a two-part lesson. In this lesson children learn about the life cycle stage of puberty in girls. By the end of this lesson they will know about some of the physical changes involved. In Lesson 8 children learn about puberty in boys. These lessons can be taught to mixed or single gender groups but all children should learn about changes in boys and girls. This lesson can be taught to boys after Lesson 8 if you wish to first teach single gender groups about the changes that will happen to them.

This lesson covers only the science aspects of a Sex and Relationships Education (SRE) programme; some questions and issues arising from it may be more appropriately addressed in PSHE lessons. Reference should be made to the school's SRE policy.

Key vocabulary:

puberty, reproduction, genitals, vagina, pubic hair, underarm hair, menstruation, period, eggs, breasts, hips, grow, shape, sweat, hygiene, spots, mood

Resources:

Large sheets of paper and coloured pens or pencils for poster making; video camera, tablet computer with camera, or sound recording equipment, if available

Health and safety:

Remind children that they should not disclose publically sensitive information about themselves or others. If they are worried by anything in the lesson they should discuss it privately with an appropriate adult. Children who do share information must be told that you cannot guarantee to keep it confidential; if a child discloses something of concern it must be reported to the member of staff responsible for child protection.

National curriculum links:

Describe the changes as humans develop to old age

Learning intention:

To describe puberty in girls

Scientific enquiry type:

Grouping and classifying

Working scientifically links:

Reporting and presenting findings from enquiries, including conclusions, causal relationships and explanations of and degree of trust in results, in oral and written forms such as displays and other presentations

Success criteria:

- I can identify differences between girls and women.
- I can describe the changes that happen to girls during puberty.
- I can give reasons for some of the changes.
- If this lesson is taught after Lesson 8:
  I can compare puberty in males and females.

EXPLORE:

Share the learning intention and explain to children that they will be learning about puberty in more detail. They are considering boys and girls separately; this lesson is about girls. Remind children about the ground rules that are used in PSHE lessons, including the use of correct vocabulary where known, rather than family or playground words. Provide each pair of children with the pictures of girls and women (Resource sheet 1), a sheet of paper and a pen. Ask them to sort the pictures into two groups – women and girls – and annotate the group or individual pictures within it to show what characteristics each group has. Remind the children that some of them may not be visible in the pictures. Take feedback from the pairs, list the features that have been identified and challenge any misconceptions. The outcomes from this activity will help you to identify children who may need additional support during the challenge activities. Children may already be familiar with some changes if your SRE programme introduces them in earlier year groups.

ENQUIRE:

Show and discuss the animation (Animation 1). Ensure that the discussion includes the answers to relevant questions from Lesson 6. Follow school policy for dealing with questions about content not included in the Key Stage 2 science programme of study; you may need to tell children that they will learn this when they are older or that it will be covered in PSHE. Note the questions and put them in the question box.

The children work in pairs to complete the challenges. They are differentiated by the method of information presentation.

The challenges are presented on the Challenge slides to be displayed on the board, or printed out and placed in the centre of the table.

Challenge 1: Children present information and advice about puberty in girls on an annotated diagram

The children in this group will use an enlarged version of Resource sheet 2. They annotate it to show the main physical changes, including information about the significance of the changes, such as breasts produce milk for babies, and any useful advice, such as about personal hygiene.

Challenge 2: Children present information and advice about puberty in girls as an interview between a Year 5 child and an expert

The children in this group will take on the role of either a Year 5 child or an expert, such as a school nurse. They will make notes of key questions and the important information that will be given in the answers, then rehearse it as an interview between the child and the expert. If suitable equipment is available the children could record or video their interviews.

Challenge 3: Children present information and advice about puberty in girls as an information poster

The children in this group will produce a poster with the title 'Ten things girls need to know by the age of 10', for display in the waiting room of a clinic or surgery.

Ask: *What is the reason girls need to change in this way when they become women? Which of these changes usually happen first/last? Are you using the correct scientific words and explaining what they mean?*

REFLECT AND REVIEW:

If teaching this lesson before Lesson 8: Ask some or all of the children in Challenge group 2 to perform their interviews.

Ask: *What was good about the interview? Was all the important information included? Were the explanations clear? What could have been improved?*

If teaching this lesson after Lesson 8: Choose only one interview for performing and evaluating. Then ask children to consider what they have learned about puberty in boys and girls and to discuss with a partner which changes are specific to boys or girls and which happen to both. Summarise the discussions on a Venn diagram. Remind children that there are similarities and differences between puberty in individual boys and individual girls as well as between boys and girls.

Remind children that now, or later when they have had time to think about it, they can add further questions to the question box. You will need to find a time (probably in a related PSHE lesson or circle time) to answer them.

EVIDENCE OF LEARNING:

Can children identify the main visible differences between the girl and the woman? Can they give differences which are less obvious? Do they know the main changes of puberty in girls? Are they using the correct scientific vocabulary for the body parts and changes? Can they describe puberty as a series of changes happening over several years? Do they recognise that everyone is different: the changes of puberty happen at different times and some do not happen in the same way to everyone?

If this is taught as the second lesson about puberty: Can they recognise similarities and differences between puberty in boys and girls, including changes such as increase in body hair and deepening voice, which happen to both but are more noticeable in boys?

CROSS CURRICULAR OPPORTUNITIES:

This lesson links with the school's PSHE scheme of work and should be taught in conjunction with the sex and relationships element of this.

REPRODUCTION IN PLANTS AND ANIMALS

LESSON 8: HOW DO BOYS BECOME MEN?

LESSON SUMMARY:

This is the second of a two-part lesson. In this lesson children learn about the life cycle stage of puberty in boys. By the end of this lesson they will know about some of the physical changes involved. In Lesson 7 children learn about puberty in girls. These lessons can be taught to mixed or single gender groups but all children should learn about changes in boys and girls. This lesson can be taught to boys before Lesson 7 if you wish to first teach single gender groups about the changes that will happen to them.

This lesson covers only the science aspects of a Sex and Relationships Education (SRE) programme; some questions and issues arising from it may be more appropriately addressed in PSHE lessons. Reference should be made to the school's SRE policy.

Key vocabulary:

puberty, reproduction, genitals, penis, testicles, sperm, pubic hair, body hair, underarm hair, facial hair, larynx (Adam's apple), voice breaking, grow, shape, perspiration, hygiene, spots, mood, muscles

Resources:

Large sheets of paper and pens

Health and safety:

Remind children that they should not disclose publicly sensitive information about themselves or others. If they are worried by anything in the lesson or by anything they are asked to do by an adult or older child they should discuss it privately with an appropriate adult. Children who do share information must be told that you cannot guarantee to keep it confidential; if a child discloses something of concern it must be reported to the member of staff responsible for child protection.

National curriculum links:

Describe the changes as humans develop to old age

Learning intention:

To describe puberty in boys

Scientific enquiry type:

Grouping and classifying

Working scientifically links:

Reporting and presenting findings from enquiries, including conclusions, causal relationships and explanations of and degree of trust in results, in oral and written forms such as displays and other presentations

Success criteria:

- I can identify differences between boys and men.
- I can describe the changes that happen to boys during puberty.
- I can give reasons for some of the changes.
- If this lesson is taught after Lesson 7: I can compare puberty in males and females.

EXPLORE:

Share the learning intention and explain to children that they will be learning about puberty in more detail. They are considering boys and girls separately; this lesson is about boys. Remind children about the ground rules that are used in PSHE lessons, including the use of correct vocabulary where known, rather than family or playground words. Provide each pair of children with the pictures of boys and men (Resource sheet 1), a sheet of paper and a pen. Ask them to sort the pictures into two groups – men and boys – and annotate the group or individual pictures within it to show what characteristics each group has. Take feedback from the pairs, list the features that have been identified and challenge any misconceptions. The outcomes from this activity can inform the grouping of children for the challenge activities. Children may already be familiar with some changes if your SRE programme introduces them in earlier year groups.

ENQUIRE:

Show and discuss the animation (Animation 1). Ensure that the discussion includes the answers to relevant questions from Lesson 6. Follow school policy for dealing with questions about content not included in the Key Stage 2 science programme of study; you may need to tell children that they will learn this when they are older or that it will be covered in PSHE. Note the questions and put them in the question box.

The challenges are differentiated by the amount of structure they provide for describing the changes of puberty.

The challenges are presented on the Challenge slides to be displayed on the board, or printed out and placed in the centre of the table.

Challenge 1: Children identify changes that happen to boys during puberty

These children work in groups of two or three. Provide them with an enlarged copy of Resource sheet 2. Ask them to draw lines between the two pictures to show where changes of puberty have happened. They then write along the line to name or describe the change.

Ask: *How are the man and the boy different? What is the scientific name for this?*

Challenge 2: Children annotate a drawing to show the changes that happen to boys during puberty

These children work in groups of two or three. Provide them with the Boy changes cards (Resource sheet 3) and ask them to sort them into those that are always true and those that are sometimes true. Each child will then label the body outline (Resource sheet 4) with the changes, adding any other information they think is necessary.

Ask: *Does everyone change in the same way and at the same time? Where on the body does this change go? Can you remember anything else about this?*

Challenge 3: Children describe the process of puberty in boys

Provide the children with the Email from an alien (Resource sheet 5). They work individually or in pairs to write a description of puberty for the alien.

Ask: *Have you described all the changes? Have you explained all the scientific words you have used?*

REFLECT AND REVIEW:

If teaching this lesson before Lesson 7: Review the card sort which is part of Challenge 2, emphasising that there are similarities and differences in the way that boys change during puberty and that this is normal.

Ask children who completed the other challenges: *Did you include all the changes that are on these cards? Did you include anything else?*

If teaching this lesson after Lesson 7: Ask children to consider what they have learned about puberty in boys and girls and to discuss with a partner which changes are specific to boys or girls and which happen to both. Summarise the discussions on a Venn diagram. Remind children that there are similarities and differences between puberty in individual boys and individual girls as well as between boys and girls.

Remind children that now, or later when they have had time to think about it, they can add further questions to the question box. You will need to find a time (probably in a related PSHE lesson or circle time) to answer them.

EVIDENCE OF LEARNING:

Can children identify the main visible differences between the boy and the man? Can they give differences which are less obvious? Do they know the main changes of puberty in boys? Are they using the correct scientific vocabulary for the body parts and changes? Can they describe puberty as a series of changes happening over several years? Do they recognise that everyone is different: the changes of puberty happen at different times and some do not happen in the same way to everyone?

If this is taught as the second lesson about puberty: Can they recognise similarities and differences between puberty in boys and girls, including changes such as increase in body hair and deepening voice, which happen to both but are more noticeable in boys?

CROSS-CURRICULAR OPPORTUNITIES:

This lesson links with the school's PSHE scheme of work and should be taught in conjunction with the sex and relationships element of this.

GET SORTED

INTRODUCTION

In this module children identify, compare and classify a variety of materials according to both their properties and their uses. They explore familiar materials in a wide range of contexts and begin to recognise that a single material name, like 'metal' or 'plastic' can describe a considerable number of different materials that may display very different properties, but which still have features in common. Specific scientific and other vocabulary is used by children as they describe, explain and communicate their understanding of materials, succinctly and in ways appropriate to a science context.

Please note that solubility, although included in the wide-reaching NC statement 'Compare and group together everyday materials based on evidence from comparative and fair tests, including their hardness, solubility, transparency, conductivity (electrical and thermal), and response to magnets' is not explored in this module, as it is covered in depth in Module 5.

When working scientifically, children plan and carry out different enquiry types to answer questions, including their own, about materials and their uses. They sort, compare, group and classify materials, and develop their abilities to plan and carry out comparative and fair tests, controlling variables, as appropriate.

This module, together with Modules 4, 5 and 6, builds on earlier learning that began in Key Stage 1 and then continued in Year 4. During those years children compared and grouped materials according to whether they were solids, liquids or gases, and learned about changes of state that take place when materials are heated or cooled. This series of modules offers the final chemistry-related learning for children in Key Stage 2. It provides teachers with ample opportunities to assess children's progress against the programmes of study and in readiness for end of key stage assessment.

Key vocabulary:

properties, material, solid, liquid, gas, compare, contrast, group, organise, criteria, hardness, soluble, insoluble, transparent, transparency, opaque, hardness, strength, rigidity, flexibility, elastic, elasticity, ductile, electrical conductor/insulator, thermal conductor/insulator, magnetic, non-magnetic, attract, repel, viscosity, viscous, thick, thicker, types of plastic – polyester, nylon, polythene, PVC, polystyrene acrylic – recycle, reuse, biodegradable, environmentally friendly

FACT FILE:

Materials and their properties

Children need to have experience of and explore as many different materials as possible in order to make sense of their world. Understanding how materials behave in their natural state and under certain conditions will help them to understand that objects are made of specific materials because of the properties of those materials. Some properties are easily observable, such as transparency, which they explore as younger children, while others are less obvious and need to have tests carried out on them in order for understanding to develop.

In Year 4, Module 1, In a State, children will have encountered materials in three states – solid, liquid and gas – and explored the properties of those different states. They learned that:

• **Solids** retain their shape when transferred from place to place unless a force is applied to them, for example, to cut or shape them. They have constant volume. This is because the particles making up the solid are held in a tight structure where they can vibrate but cannot move in relation to each other. Powders can be poured but will form a pile rather than a pool (flat surface). Each grain of a powder maintains its shape and volume.

• **Liquids** when transferred from place to place take the shape of the container they are in but do not change in volume. The surface of a liquid will remain horizontal when the container is tipped. The particles in a liquid remain in contact with each other so the liquid cannot be compressed, but they are more loosely bound and so can move in relation to each other, allowing changes of shape.

• **Gases** change in shape and volume to fill the space they are in. The particles in a gas move freely so, under pressure, the gas will take up less space.

During this module children build on this knowledge as they develop their understanding of materials further. They explore a variety of solid materials of different densities, establishing a continuum of solidity as they explore how soft and hard it is possible for a solid to be.

Conductors and insulators

A conductor is a material that transmits something like electricity or heat well. An insulator is a material that does this less well or not at all. Metals are very effective thermal and electrical conductors. Plastics and woods are poor thermal and electrical conductors but very good thermal and electrical insulators. Both these materials are ideal for using in contexts where heat requires insulating to protect the user, for example, for the handles of metal cooking pans or for cooking spoons that are used with hot food. However, wood may char or burn if it becomes too hot and plastic might melt – so care does need to be taken. Plastics are also used to insulate cables and plugs on electrical appliances and in the wiring of our homes, specifically because they are such effective electrical insulators.

Unusual materials and how they behave

Many of the materials children encounter during this module behave as they might expect, based on prior experience. But some materials may surprise children, behaving differently and challenging their expectations. Most of these materials, for example, lemonade, ketchup and jelly, will be encountered frequently by children during their everyday lives. They are familiar materials, but the lessons within this module require children to think about and respond to them in a different way.

Some examples of these materials, and simple explanations of their behaviour, are included below.

Shaving foam: The combination of glycerine, lanolin and other chemicals gives shaving foam its extra-creamy and dense lather. It combines with a propellant (often butane or propane gas), which expands and instantly evaporates when it leaves the can, filling the foam with millions of bubbles.

Lemonade: Compressed carbon dioxide is forced into the still drink during production. When the pressure of gas inside is released, as the lid is opened, the carbon dioxide rushes to leave the liquid to equalise the pressure.

Jelly: Jelly contains gelatin, which changes state when mixed with hot water. On cooling, it sets into a semi-solid flexible mass.

Ketchup: Tomato ketchup is made from concentrated tomatoes. It is an example of a non-Newtonian fluid – liquids that get thinner or thicker when 'stress' is exerted. In the case of ketchup, it stays almost solid in the bottom of a plastic bottle until it is squeezed or shaken, at which point it squirts out easily: it is 'shear thinning'.

Cornflour: Cornflour mixed with water is another example of a non-Newtonian fluid – in this case one that is 'sheer thickening', i.e. it becomes more solid when under stress. When water is added the large starch particles that make up cornflour become suspended, move slowly around and behave like a liquid. The faster the mixture is stirred the more viscous (thick) the material becomes. This is because at low speeds the water can easily fill gaps between particles, but at higher speeds the water is unable to do so. As friction increases, the viscosity increases! In a Newtonian fluid (like water), viscosity remains constant as the fluid is stirred.

Common misconceptions:

Children sometimes use the word 'material' to describe fabric and textiles. They need to be reminded that in science the word 'material' is a generic adjective used to describe what something is made of.

Many children believe that all metals are magnetic. Only metals containing iron (including steel), nickel and cobalt are magnetic (i.e. can be attracted to a magnet).

GET SORTED

LESSON 1: HOW CAN WE COMPARE AND GROUP MATERIALS?

Key vocabulary:

properties, material, compare, contrast, group, organise, criteria, hardness, soluble, insoluble, transparent, opaque, electrical conductor/insulator, thermal conductor/insulator, magnetic, non-magnetic, attract, repel

Resources:

Sticky notes, large sheets of paper, familiar classroom objects, for example, marker pen, pencil, paper clip, plant pot, sweatshirt, sports shoe, stapler, ruler, water bottle, lunch box, eraser; real objects and substances, for example, milk, shaving foam, ketchup, butter, yoghurt, jelly, hair gel, steam, sand, flour, sugar

Key information:

Children sometimes use the word 'material' to describe fabric and textiles. They need to be reminded that in science the word 'material' is a generic term used to describe what something is made of. Some materials have easily observable features, such as transparency, whereas the features of others are less obvious and need to be identified using a range of tests.

LESSON SUMMARY:

In this lesson children identify, compare and group materials based on their properties and according to their own or given criteria. They build on learning started in Key Stage 1 and continued in Year 4, Module 1, In a State. Children use a variety of ways to do this, including creating keys in some cases. By the end of this lesson children will be able to describe the properties of materials in more depth, identifying more specific and testable properties of familiar and less familiar materials.

National curriculum links:

Compare and group together everyday materials based on evidence from comparative and fair tests, including hardness, solubility, transparency, conductivity (electrical and thermal) and response to magnets

Learning intention:

To classify a variety of materials according to their properties

Scientific enquiry type:

Grouping and classifying

Working scientifically links:

Recording data and results of increasing complexity using scientific diagrams and labels, classification keys, tables, scatter graphs, and bar and line graphs

Success criteria:

- I can make comparisons between different materials, using technical vocabulary to accurately describe their properties.
- I can identify specific criteria to help me compare and group materials.
- I can create a key to help me classify different materials.

EXPLORE:

Organise children into groups and give each a selection of familiar classroom objects to discuss, for example, a marker pen, pencil, paper clip, plant pot, sweatshirt, sports shoe, stapler, ruler, water bottle, lunch box, eraser. Make sure that there is a good mix of objects made of different materials and with different properties. Check that children know what all the objects are.

Ask: *What materials are the objects made from? What properties could describe the materials?*

Make sure that the distinction is established here between the materials the objects are made from, for example, metal, plastic, and the properties of those materials, for example, hard, shiny. Ask the children, in groups, to write the property words that they have come up with on separate sticky notes. Take feedback from the children about the properties they have identified. Prompt them to use technical vocabulary to describe properties, for example, flexible, rigid, transparent, translucent, opaque, conductor, insulator, (electrical and thermal), magnetic.

Explain to the children that they are going to play 'materials dominoes' in their groups.

Model the game yourself first. Place an object where the children can see it. Identify two material or property words for that object. For example, for a ruler, the material and property words could be plastic, flexible. The children then have to match either the property (two points) or material (one point). Take several suggestions, identifying two words and making a match each time. The children should then try with the objects on their tables, aiming to use all of them and score the maximum number of points.

ENQUIRE:

Use your observations of children's responses to the Explore part of the lesson to help you group children for the challenges. Challenges are differentiated according the support given and the independence children have to determine their own classification system.

The challenges are presented on the Challenge slides to be displayed on the board, or printed out and placed in the centre of the table.

Challenge 1: Children sort a collection of materials: first into solids, liquids and gases and then using their own criteria

Give the children a collection of real materials to handle and explore. Comparing materials (Resource sheet 1) includes a list of 15 possible materials and objects. Choose an appropriate selection from them if you feel 15 is too many for the children to compare. Ask the children to use the grid to classify their collection.They should begin with solid, liquid and gas as sort criteria to establish their baseline understanding of this core knowledge (taught in Year 4). They should then move on to identifying their own criteria.

Ask: *Are all the materials that you have said are solids the same? What property could describe the difference/s? Are any of the materials transparent? Would any conduct electricity?*

Choose three of the materials and ask the children to discuss what is similar about them.

Ask: *Are there any other materials in your selection that have the same properties? Are there any other materials that you can think of with these properties?*

Challenge 2: Children compare and group a selection of materials according to their own criteria

Give the children a selection of materials, for example, milk, shaving foam, ketchup, butter, yoghurt, jelly, hair gel, sand, flour, sugar.

Be ready to offer prompts to get them started, and refer them to the Property definition list (Resource sheet 2) if necessary. As they think about the criteria they will choose to compare and sort materials, challenge the children to think about less obvious or visible properties, for example, whether the material/s would be waterproof, or soluble in water, or magnetic.

Ask: *How are the materials similar? How are they different? Which properties have you used to sort your materials? Is there a property that is only appropriate to one or two materials?*

Challenge 3: Children create a key to sort a selection of materials

Give the children a selection of materials, for example, milk, shaving foam, ketchup, butter, yoghurt, jelly, hair gel, sand, flour, sugar.

Provide them with an enlarged version of the generic blank key or large sheets of paper plus sticky notes, so that they can plan their questions to sort their materials and develop their key. Remind the children that they need to think of questions that can be answered yes or no. When they have completed their key ask them to test it out on a friend.

Ask: *How easy did you find it to sort your selection of materials completely? Was your friend able to answer your questions and use them to sort the materials? Could you ask different questions and come up with another key that would still separate them out completely?*

REFLECT AND REVIEW:

Provide a selection of objects made of different materials (including some that have already been investigated) for children to compare and contrast using Resource sheet 3. Encourage them to suggest which objects they might compare.

Ask: *How are the two materials you have chosen the same? Which properties do they have in common? How are they different?*

Select and show children two objects that are made of the same material in different forms, for example, a metal food tray and a rigid metal object, like the frame of a chair. Again, ask children about the similarities and differences between the materials. Ensure that children recognise that both are solids and both are made of metal, and that they share some properties but also have different properties, for example, the food tray is much less rigid than the chair frame.

EVIDENCE OF LEARNING:

Observe the children working and listen to their thinking and ideas as they compare and group materials:

• How effectively do they make comparisons between different materials, identifying their similarities and differences?

• Do they use technical vocabulary accurately to describe the properties of materials?

• Can they identify specific criteria to use as they compare, sort and group materials? Are some of these criteria more specific, revealing a deeper understanding of the properties of those materials?

• Were they able to construct a key to sort a collection of materials?

GET SORTED

LESSON 2: IS A SOLID ALWAYS HARD?

LESSON SUMMARY:

In this lesson children investigate solids and compare them according to their properties. By the end of the lesson they will be able to identify a range of soft to hard solids and sequence them according to the property of hardness.

Key vocabulary:

properties, material, compare, contrast, group, organise, sequence, criteria, hard, hardness, transparent, transparency, malleable, malleability, elastic, elasticity, flexible, flexibility, brittle, permeable, impermeable, permeability, conductor, insulator, solid

Resources:

Microscope, marshmallows and jelly sweets, chocolate buttons, cheese strings, cooked pasta, foil, elastic, net (or old tights), sponge, polystyrene, sand, soil, butter, brick, wooden ruler, plastic toy, metal object, piece of fabric, glass bottle, sponge, corn flour, water, tray or large bowl

National curriculum links:

Compare and group together everyday materials based on evidence from comparative and fair tests, including hardness, solubility, transparency, conductivity (electrical and thermal) and response to magnets

Learning intention:

To compare and contrast different solids according to their properties, including their hardness

Scientific enquiry type:

Carrying out comparative and fair tests

Working scientifically links:

Reporting and presenting findings from enquiries, including conclusions, causal relationships and explanations of and degree of trust in results in oral and written forms such as displays and other presentations

Success criteria:

- I can compare the properties of solid materials.
- I can plan a test to group and classify solids according to their hardness.
- I can sequence a range of solid materials according to the property of hardness.
- I can describe how the hardness of solids differs.
- I can explain what is different about the structure of a soft and a hard solid.

EXPLORE:

Show children Slideshow 1, which shows comments from children talking about a collection of materials and sharing their points of view.

Ask: *Which of these statements do you agree with? Which do you disagree with? Which are you not sure about it? Can you give reasons why?*

ENQUIRE:

Organise children into groups and explain that they are going to investigate solids to find out how soft and hard it is possible for a solid to be. The three challenges describe a sequence of learning that might be attempted by all of the children, with an appropriate level of support.

The challenges are presented on the Challenge slides to be displayed on the board, or printed out and placed in the centre of the table.

Challenge 1: Children sort objects according to their properties and then test them to order them according to softness

Ask: *How soft can a solid be, and still be classified as a solid? Are all solids soft in the same way? What do you mean by the word 'soft' when you are describing a solid?*

Provide the children with a selection of solids to explore, for example, marshmallows and jelly sweets, chocolate buttons, cheese strings, cooked pasta, foil, elastic, net (or old tights), polystyrene, sand, soil. Ask the children to group them first, thinking about their properties, identifying their similarities and using these similar properties to group them.

Ask: *How could you test the materials to decide which is the softest?*

Encourage the children to carry out a simple comparative test to establish a rough order of softness across the materials in the selection. If necessary, model ways in which they might compare the solids, for example, by observing how much they stretch when pulled, or how much they squash when a mass is placed on top of them. The children should write a brief description of how they carry out the test and what they find out.

Key information:

Children will have learned about the concept of states of matter, i.e. solids, liquids and gases, in Year 4. Check that they remember the definitions. Some children may still struggle to recognise that a 'soft' solid, like sponge or foam (used in soft furnishing), is a solid, as are sand and flour, which behave a little like liquids when spilt from a container. Further discussion and practical exploration might be necessary. Provide children with access to a microscope to examine sand, for example. This will allow them to look closely at grains of sand and see the solid structure of each one.

Challenge 2: Children compare and contrast objects and place them on a hardness scale

Provide the children with a block of butter and a brick to compare (and other objects if possible) and a copy of the Compare and contrast grid (Resource sheet 1), and the Hardness scale (Resource sheet 2). Explain to the children that they are going to use the questions to compare these two solids, describing their observations in as much detail as possible.

Ask: *On a hardness scale from 1 to 10 (10 being the hardest), where would butter and a brick be? Can you identify solids that you predict will be harder than brick, softer than butter or anywhere in between?*

The children should record their predictions on the scale, showing where they think different solids should be placed.

Ask: *How can you test these solids to make sure you have put them in the right place on the scale?*

Challenge 3: Children compare solids and explain how hard and soft solids differ

Provide the children with a selection of solids representing a full range of solid materials, for example, sand, wooden ruler, plastic toy, rice, aluminium foil, rigid metal object, piece of fabric, glass bottle, soil. Ask the children to select three of the materials and draw an annotated diagram of how they think the particles in these solid materials are arranged.

Ask: *What do you think might be different about the particles in a soft solid and a hard solid? How do you think they are arranged? Remember in Year 4, when you observed butter melting? What happened to the particles in the solid? How do the particles move as melting takes place?*

REFLECT AND REVIEW:

Return to the image that children discussed during the Explore part of the lesson (Slideshow 1).

Ask: *Do you remember what you thought at the beginning of the lesson? Have you changed your mind at all? Why is that? What have you found out? Could you come up with a definition that might help other children understand that solids can be very different – but still solids?*

EVIDENCE OF LEARNING:

Listen to children's responses throughout the lesson. Can they compare a variety of solid materials according to their properties? Can they rank a variety of solid materials according to the property of hardness? Can they describe the different properties of soft solids and hard solids in detail, naming examples and comparing them? Are some children able to describe why the hardness of solids differs?

GET SORTED

LESSON 3: IS A LIQUID ALWAYS RUNNY?

LESSON SUMMARY:

In this lesson children carry out various comparative tests, exploring the viscosity of liquids. By the end of this lesson they will have completed a comparative test, used the evidence collected to order liquids from the thinnest to the thickest, and be able to describe how viscosity varies from liquid to liquid.

Key vocabulary:

properties, material, compare, contrast, viscosity, viscous, transparent, transparency, liquid, pour, flow

Resources:

Large sheets of paper, honey, cooking oil, syrup, milk, washing up liquid, bubble bath, lemonade, yoghurt, different brands of tomato ketchup, wipe-clean ramps, whiteboards, teaspoons, tablespoons, stop watches or watches with second hands

Health and safety:

Ensure that all liquids are safe for children to use.

National curriculum links:

Compare and group together everyday materials based on evidence from comparative and fair tests, including their hardness, solubility, transparency, conductivity (electrical and thermal) and response to magnets

Learning intention:

To compare and contrast the properties of different liquids, including viscosity

Scientific enquiry type:

Carrying out comparative and fair tests

Working scientifically links:

Planning different types of enquiries to answer questions, including recognising and controlling variables where necessary

Success criteria:

- I can test the properties of liquids and compare them.
- I can plan a test to compare the viscosity of different liquids.
- I can sequence a variety of liquids according to how viscous they are.

EXPLORE:

What is a liquid?

Print out the Concept words (Resource sheet 1) on card, cut them out and give each table of children a set of the cards, together with a large sheet of paper and pens. Explain that they are going to create a concept map using these words, linking the words with arrows and labelling the links. If necessary, model this process quickly using two or three of the concept words, adding links and defining the link, for example, washing up liquid and oil both linked to container; link phrase 'can be held in'.

Check that they all have a clear understanding of the difference between solids that behave like liquids and liquids themselves.

ENQUIRE:

Remind children of the question that is the focus of the lesson: Is a liquid always runny?

Ask: *Is that true? How would you describe a liquid? Are all liquids the same?*

Ensure that children identify relative thickness as one of the ways in which liquids differ. Introduce the terms 'viscous' (thick) and 'viscosity' (how thick something is).

Explain that during this lesson they are going to be finding out about the viscosity of liquids. In Challenges 1 and 2 children will be planning and carrying out a comparative test to discover which liquid is the thickest. Those completing Challenge 3 will be product testing different brands of ketchup to discover whether they are all the same.

The challenges are presented on the Challenge slides to be displayed on the board, or printed out and placed in the centre of the table.

Challenge 1: Children carry out a comparative test to find out which is the thickest liquid

Organise children into groups and provide each group with a set of liquids (alternatively you could organise this as a carousel activity, placing a different liquid on each table and asking the groups to move around as they test them).

Show them a test method they might use, for example, timing how long it takes for a liquid to pour from an unshaken bottle or from a large spoon, or timing how long after pouring it takes for a quantity of liquid to form a flat surface.

Ask the children to record their results in a bar chart using the generic graph template, and to label the axes with the different type of liquid (x axis) and time taken (y axis).

Key information:

Cornflour is powdered white starch extracted from maize kernels. It exhibits the properties of a shear thickening fluid (also called a non-Newtonian or dilatant fluid).

Cornflour has large particles. When you add water, the particles become suspended, move slowly around and behave like a liquid. The faster you stir the more viscous (thick) the material becomes. This is because at low speeds the water can easily fill gaps between particles, but at higher speeds the water is unable to do so. As friction increases, the viscosity increases. Tomato ketchup, on the other hand, is a shear thinning fluid because its viscosity decreases as it is shaken, which also makes it a non-Newtonian fluid. In a Newtonian fluid viscosity remains constant as the fluid is stirred.

After they have completed their investigation, ask them to order the liquids from thinnest to thickest.

Ask: *Can you think of any other thick liquids we could have tested? Where would they fit into the order?*

Challenge 2: Children plan a comparative test to find out which is the thickest liquid, including identifying a method, setting up equipment, and recording what they find

Encourage the children to think of ways in which they could find out an answer to the question: Which is the thickest liquid? If necessary prompt them by offering a box of materials that they might test. They could shake and listen or look at how the liquids move in their containers to identify an order from thinnest to thickest.

Ask: *How do we know which of these is the thickest? How could you measure more accurately the thickness of each liquid?*

If the children need some help, suggest they create a trickle test, timing how long it takes for a small amount of the liquid to travel down a slope from a fixed height. The one that takes the longest to trickle down will be the thickest.

Remind the children that they will need to plan how they will set up their test and decide on the equipment they will use, what evidence they will collect and how to record their results.

Challenge 3: Are all ketchups the same? Children plan a comparative test to find out if different brands of ketchup are similar in thickness

Prompt the children to think about what they are investigating.

Ask: *Are all ketchups the same? How are they different? How does their thickness vary? Do thicker ketchups contain the most tomatoes? Are the thicker ketchups the most expensive? Why? Are some brands of ketchup redder than others? Why?*

Encourage the children to brainstorm their own questions related to the viscosity of the ketchups and then decide which they would like to investigate. They should decide on a method, what equipment to use, what evidence to collect, where they might look for that evidence and how to record it.

REFLECT AND REVIEW:

Ask children to describe to their class the comparative test they carried out and what they found out about the viscosity of liquids. Which of the liquids were the thickest? Were there any surprises?

Ask those children that completed Challenge 3: *What were the differences between the thickness of the ketchups you looked at? Did any of the ketchups include 'thickeners' in the ingredients?* Explain that this may mean they include cornflour, or a similar material in their ingredients.

Cornflour is a common 'thickener'. Show the children what happens to cornflour when water is added. Sprinkle cornflour onto a large tray. Gradually add water to the cornflour, stirring it slowly to combine the two.

Ask: *What do you notice happens when the mixture is stirred slowly? What happens when it is stirred quickly? Can you scoop some of the mixture out and roll it into a ball? What happens when you stop rolling it? What seems to make the cornflour suddenly harden?*

Explain that it is the movement of the cornflour as it is stirred that causes it to stiffen. Use the observations about the behaviour of cornflour to encourage children to think again about the viscosity of tomato ketchup.

Ask: *What do you think would happen if cornflour were added to tomato ketchup? Would the liquid ketchup become thicker? Or thinner? What do you think would happen if the ketchup were shaken up (which we often do to make it easier to pour)?*

Thickeners are sometimes used in cheap ketchups. Good ketchups are thick because they include large quantities of tomatoes. These products thin when shaken, rather than thicken.

Challenge children to see what else they can find out about the science of cornflour as a homework activity.

EVIDENCE OF LEARNING:

Listen to children's responses throughout the lesson. Can they test the viscosity of a liquid? Do they independently think of a test to compare the liquids? Do they plan a method, carry out their test using equipment and record the evidence that they have generated? Can they sequence a variety of liquids according to how thick or thin they are, describing how viscosity varies from liquid to liquid?

GET SORTED

LESSON 4: ARE ALL METALS THE SAME?

LESSON SUMMARY:

In this lesson children explore the ways in which metals are used around their school and in the wider world, and link these uses to the properties of the metals. By the end of this lesson children will have identified where, how and why metals are used and explained why properties of certain metals make them especially suitable for particular purposes.

Key vocabulary:

properties, material, compare, contrast, hardness, strength, rigidity, flexibility, ductile (can be drawn into wires), electrical conductor, thermal conductor, magnetic, non-magnetic, attract, repel

Resources:

Magnets, examples of objects made of metals, for example, cooking pan, spoon, bell, paper clips, stepladder, power cable, access to books or the internet for research

Key information:

Many children believe that all metals are magnetic. Only metals containing iron (including steel), nickel and cobalt are magnetic. All metals conduct electricity and they are very effective thermal conductors.

National curriculum links:

Compare and group together everyday materials based on evidence from comparative and fair tests, including their hardness, solubility, transparency, conductivity (electrical and thermal) and response to magnets

Learning intention:

To identify the properties of different metals and describe how these properties make them suitable for particular uses

Scientific enquiry type:

Grouping and classifying

Working scientifically links:

Identifying scientific evidence that has been used to support or refute ideas

Success criteria:

- I can identify the properties of a variety of different metals.
- I can link the properties of metals to how they are used.
- I can explain that particular metals are used for specific purposes because of their properties.

EXPLORE:

What do we know about metals already?

Print out the Sorting cards (Resource sheet 1), cut them out and give a set to children in pairs. Ask the children to sort the statements on the cards identifying whether they Agree, Disagree, or are Not sure. When they have made an initial sort, ask them to describe how they have sorted the cards, to share their reasons and to explain why they placed any cards in the Not sure group. Use this activity as an opportunity to clarify children's understanding of metals and their properties, and to identify any misconceptions that they may have and opportunities for further teaching later in this and other lessons.

ENQUIRE:

Explain to children that during this lesson they will be finding out more about properties of metals and what metals are used for.

The challenges are differentiated through support as well as through the science knowledge required to complete each of them. Use the outcomes of the Explore card sort to help you decide which challenge will be most appropriate for children to carry out. Challenge 1 requires children to find evidence of how metals are used and a grid structures their investigation. Challenge 2 requires children to give reasons why particular materials are used for specific purposes, while Challenge 3 asks children to use existing knowledge and research to construct an evidence-based argument.

The challenges are presented on the Challenge slides to be displayed on the board, or printed out and placed in the centre of the table.

Challenge 1: Children identify ways in which metals are used around school and, from research, link the uses of different metals to their particular properties

Provide the children with the Properties of metals table (Resource sheet 2). Explain that their challenge is to identify at least one example of a metal being used because it has a particular property on the list. Examples might be school-based or from the wider world. If allowed, encourage children to move round the public areas of the school to look for other examples of metallic objects.

Ask: *Which property was the easiest to find examples of? Which metals are used and for what purposes? Why do you think certain metals are especially suited to particular jobs, for example, why is aluminium often used to make stepladders?*

Key Information:

Metals share a combination of generic properties to different extents, for example, they are strong, flexible, malleable, ductile, conduct electricity and heat. This makes them particularly suited for specific uses, for example, lead is softer than steel and is more easily worked at low temperatures. Copper is a particularly good conductor of electricity, which is why it is used in cables and for wiring electrical circuits.

Challenge 2: Children explore why different metals are used for particular purposes because of their properties, by finding answers to questions of their own or from a list

Explain to the children that another class has come up with some questions about metals and why they are used for certain jobs. Ask them if they can help them answer these, or think of some questions of their own to answer. Share some example questions, presenting them to the children as a list: Why is metal used to make cooking pans? Why is metal used for power cables? Why is metal used to make drawing pins, or staples? How can metal help to close the fridge door? Why might I find metal in my breakfast cereal? What if a frying pan were made of plastic? What if a cymbal were made of glass? Give them a copy of the 'I wonder why' table (Resource sheet 3) to complete, either individually or one between two.

Challenge 3: Children investigate, using secondary sources, the answer to a question and present their ideas in the form of a reply email

Give the children a copy of the Email argument (Resource sheet 4), one between two. The email discusses a difference of opinion about why foil is shiny on one side only.

Ask: *What difference do you think it makes? Where might you find evidence to answer the question? What does Joe think? And Jay? Who do you agree with?*

Explain to the children that they need to find some answers, complete with evidence, and write a group reply email to Joe and Jay. The reply needs to explain whether they are right or wrong and give the scientific explanation for the shininess of aluminium foil. Provide access to books or the internet so that the children can carry out some research and try to solve the argument.

Note: The reason aluminium foil is shiny on one side and dull on the other is nothing to do with cooking food. Shiny or dull side out makes no difference to the foil's effectiveness as a conductor of heat. It is all about how the foil is manufactured.

Aluminium foil is produced by rolling it between successive steel rollers. Each set of rollers squeezes the foil thinner. The last stage of rolling reduces the foil's thickness to a thousandth of an inch or even thinner. The problem is that such a thin foil is too easy to tear during the last rolling stage. So, for the last stage of rolling, two sheets of foil are placed face to face and passed though the final set of rollers together. Since the steel rollers are highly polished, the foil faces that contact the rollers are also embossed to a highly shiny surface.

After rolling, the resulting sheets are separated from each other. The surfaces of each foil that had faced each other are matte in texture, since they had only been squeezed against each other and not the polished rollers.

REFLECT AND REVIEW:

Revisit the 'What do we know about metals already?' discussion and card sort from the beginning of the lesson. By this stage of the lesson children should have a much clearer understanding of the properties of metals and how their properties make them suitable for particular purposes.

Ask: *Are all metals the same? How are they different?*

Show children Are all metals the same? (Interactive 1) and use it to check children's understanding. Discuss ways in which some of the individual properties differ between the metals, such as why some are harder than others (for example, compare lead with steel), and how these differences influence the ways in which metals are used. Go on to part 2 of Interactive 1 to identify how different metals are used.

EVIDENCE OF LEARNING:

Listen to children's responses throughout the lesson. Can they identify the properties of a variety of different metals, including which metals are magnetic? Can they describe how the generic properties of metals make them suitable for a wide range of uses around school and in the wider world? Are children able to explain that particular metals are used for specific purposes because of their properties, and give examples (for example, aluminium used for making super-thin foil, steel used for door closures, copper used for wires in cables)? Are they able to identify evidence that they have gathered during the course of their challenge to support or refute their ideas? Can they state how confident they are in what they have found out and talk about the quality of the secondary sources of evidence that they may have used?

GET SORTED

Key vocabulary:

properties, material, compare, contrast, hardness, strength, rigidity, flexibility, electrical insulator, thermal insulator, polyester, nylon, polythene, PVC, polystyrene, acrylic, recycle, reuse, biodegradable, environmentally friendly

Resources:

Large bowl or jug, variety of large serving spoons made out of plastic, wood or metal, collection of objects made of plastics, for example, plastic bottles and packaging, plastic jugs and bowls (polythene), clothing made of polyester, strong ropes, washing line (nylon), beakers, plates, disposable cutlery, yoghurt pots (hardened polystyrene), insulation and packaging materials (expanded polystyrene), perspex sheets, lenses in torches (acrylic), pencils of different hardness, polystyrene cup, lemonade bottle, shampoo bottle, carrier bags, cling film, dustbin, washing up bowl or classroom tray, access to the internet or books for further research

Health and safety:

Ensure that children take care around very hot water.

Key information:

Please note that some of the properties mentioned in the NC link, for example, solubility, transparency, are not directly relevant to this lesson. Learning related to these particular properties is an element within other lessons and modules.

LESSON 5: ARE ALL PLASTICS THE SAME?

LESSON SUMMARY:

In this lesson children identify and investigate the wide-ranging properties of plastics. By the end of this lesson they will have sorted, grouped and tested plastics according to their properties, recognised how extensively plastics are used by society and considered the importance of reducing the use of plastics, and their reuse and recycling.

National curriculum links:

Compare and group together everyday materials based on evidence from comparative and fair tests, including hardness, solubility, transparency, conductivity (electrical and thermal) and response to magnets

Learning intention:

To identify the properties of different plastics and explain how these make them suitable for particular purposes

Scientific enquiry type:

Grouping and classifying

Working scientifically links:

Planning different types of scientific enquiries to answer questions, including recognising and controlling variables where necessary

Success criteria:

- I can identify different plastics and their properties.
- I can link these properties to how they are used.
- I can carry out an enquiry to answer a question about plastics.
- I can describe how the number of plastics used might be reduced and the importance of recycling and reusing plastics.

EXPLORE:

Show children a selection of large spoons made of wood, plastic or metal.

Ask: *What materials are these spoons made of? When might we use them?*

Explain that you are going to put one spoon made of each material into some very hot water. Ask children what will happen to the spoons, and to say which spoon handle they think will get hotter more quickly. Invite some children to carefully feel the handles of the spoons and feed back their observations to the rest of the class. Were their predictions correct? Which spoon handle got hottest, and the most quickly?

Introduce children to, or remind them of, the terms 'thermal insulator' and 'thermal conductor'.

Ask: *Are plastics good or poor insulators?* (good). *What about metals?* (poor). *Are woods good or poor conductors?* (poor). *What about metals?* (good).

Encourage children to think about the properties of these different materials and how they might be affected by how they are used.

Ask: *When might a wooden spoon be better to use than a metal spoon? When might a plastic spoon be better than a wooden spoon?*

Explain that we also use the terms conductor and insulator when talking about electricity.

Ask: *Are woods, plastics and metals good or poor conductors of electricity?*

ENQUIRE:

Explain that during this lesson children will be working in groups to carry out a range of enquiries to find out about the properties of different plastics, and their suitability for use in everyday objects. Children will also find out about the potential impact that society's use of plastic is having on the environment.

The challenges are presented on the Challenge slides to be displayed on the board, or printed out and placed in the centre of the table.

Challenge 1: Children identify objects made of different plastics and describe the properties of those plastics that make them suitable for specific purposes

Provide each group with items that are made from different plastics and an enlarged copy of Plastic descriptions (Resource sheet 1). Explain to the children that they are going to examine the items

Key information:

Metals are very effective thermal conductors; they heat up extremely quickly. Plastics and woods are poor thermal conductors and very good thermal insulators. They are ideal for using in contexts where heat requires insulating to protect the user, for example, the handles of metal cooking pans and cooking spoons that are used with hot food. However, wood may char or burn if it becomes too hot and plastic might melt, so care does need to be taken.

Key information:

'Plastic' is a generic term used to describe a wide variety of materials with noticeably different properties. These properties make plastics suitable for a great many everyday uses and they are often used in place of other materials such as wood or metal.

Key information:

Plastics are very versatile materials and have become a familiar part of our everyday lives. Plastics are long lasting and do not decay easily, which is a property that can be very useful as it ensures that products do not need to be replaced as frequently as those made of less resilient materials. But this longevity can itself cause problems. Objects made of plastic, for example, packaging materials, do not decay when they are thrown away. They may end up in landfill or be discarded carelessly on roadsides or waste ground and remain in the environment, the soil and water for many, many years.

including handling the objects and looking carefully at any labels (many plastic objects have an embossed label pressed into the surface, including recycling symbols). They should sort the items and record their grouping by photographing the items and noting as many details as they can in their books, or by using the Recording grid (Resource sheet 2), describing the properties of the plastic that make it suitable for the purpose to which it has been put.

Challenge 2: Children carry out a 'scratch test' to establish which plastics are the hardest and softest, and why that property makes them useful for particular uses

Explain to the children that they are going to use the pencil scale to measure the relative hardness of different plastics.

A pencil scale of hardness – 7B (very soft) through to 7H (very hard) – can be used to measure the hardness of different materials. If necessary, allow the children to explore the different pencils so that they realise the differences between them. Starting with soft pencils they should then try to scratch the surface of the different plastics. Encourage the children to think about how important the hardness of each plastic is to the plastic's use. They should record their results on Resource sheet 2 including the name of object, the type of plastic and the property of the plastic that makes it useful. Finally, ask the children to sequence the plastics, from hardest to softest.

Ask: *How can you summarise what you have found out in a general statement? How can you compare the hardest and softest plastics in a statement?*

Challenge 3: Children evaluate the recycling habits of six people according to their environmental friendliness. They then work together to find out more about the issues raised and answer questions that they generate through research

Provide the children with a set of Ranking cards (Resource sheet 3) – one set between three. Ask them to read what the cards say and then rank the statements according to how 'environmentally friendly' they are, from 'thumbs up' (very), to 'thumbs down' (poor). There are no absolute right answers, but the children must be able to explain and justify why they have chosen to place a card in a certain position. The statements will promote discussion and the children should identify questions that arise and investigate them later through research, for example, how much better is using glass bottles than using plastic? Are some plastics more environmentally friendly than others? Provide the children with access to the internet, or books, to carry out further research.

Ask: *What property of plastics make them such an environmental problem?*

REFLECT AND REVIEW:

Ask children to share what they have discovered about the huge variety of plastics we use in everyday life. Ask them to write a newspaper headline to summarise one key fact they have learned about plastics and how they found out about that fact.

Those children who completed Challenge 3 should share what they have found as a precursor to a wider discussion of plastics and recycling.

Ask: *Did you find out which plastics can be recycled? What does the recycling symbol look like? Does that mean that all plastics marked like this will be recycled? What can we do to help?*

Explain that provision for recycling varies a lot from place to place. Their local council decides what the priorities are and makes arrangements for the disposal of plastics, alongside other waste. Children could brainstorm questions they would like to ask and then write a letter to a local councillor asking for further information.

EVIDENCE OF LEARNING:

Listen to the children's responses throughout the lesson, noting especially their understanding of thermal conductivity. Can children identify objects made of plastic? Can they explain why the properties of the plastic make it suitable for the object's uses? Can they describe how the number of plastics we use might be reduced? Can they explain the importance of recycling and reusing plastics whenever possible? Can they describe the enquiry methods they used to answer their question about plastics?

CROSS-CURRICULAR OPPORTUNITIES:

This lesson can be linked with English; for example, write a letter to a local councillor to ask about local provision for recycling. Useful information, including which plastics each authority recycles, can be found at http://www.recyclenow.com/how_is_it_recycled/plastic_bottles.html

GET SORTED

LESSON 6: TO BOUNCE OR NOT TO BOUNCE: WHY ARE SPORTS BALLS SO DIFFERENT?

LESSON SUMMARY:

In this lesson children investigate the variables that affect how a ball bounces. By the end of this lesson they will have sorted balls according to their properties and materials, planned and carried out a fair test enquiry as a group, and evaluated the success of their enquiry method and the evidence it generated.

National curriculum links:

Compare and group together everyday materials based on evidence from comparative and fair tests, including their hardness, solubility, transparency, conductivity (electrical and thermal) and response to magnets

Learning intention:

To investigate the properties of materials and their uses by planning and carrying out a fair test enquiry using different types of balls

Scientific enquiry type:

Carrying out comparative and fair tests

Working scientifically links:

Plan different types of scientific enquiries to answer questions, including recognising and controlling variables where necessary

Success criteria:

- I can sort sports balls, identifying my own criteria for sorting.
- I can identify a question and plan a fair test enquiry with my group.
- I know which variables to change and which to keep the same.
- I can relate my findings to the properties of the materials and the way in which they are used.
- I can evaluate the success of the enquiry and suggest ways that outcomes might be improved.

EXPLORE:

Show children a large variety of balls, from ping pong to rugby or cricket.

Ask: *Could you play tennis with a cricket ball or football with a rugby ball? Why not? How many different sports can you name that are played with balls? What are some differences and similarities among the balls used for different sports? What criteria could we use to group the balls?*

Ask children to talk to a partner and identify possible criteria for sorting the collection of balls. Criteria might include the material they are made from and its properties, whether they are inflated or solid, their relative hardness, or whether they bounce. Encourage them to think in more detail about how to describe the range of ball size in the collection and ask children to measure and compare the balls, where possible. Agree definitions for appropriate groupings, for example, balls with a diameter of more than 10 cm (large).

Take suggestions from children and select several criteria, beginning with size, for them to try physically. Provide them with hoops to use as criteria, and ask them to create labels for each hoop. Ropes can then be used to connect hoops if necessary, to create a large-scale key.

ENQUIRE:

How do the materials and design of a ball affect its characteristics?

Each ball is designed to be made from a specific combination of materials, in order for it to be appropriate for a particular sport. For example, a football needs to be bouncy enough to be kicked down a pitch without causing injury, and a bowling ball needs to be heavy enough to be rolled.

Explain to children that during this lesson they are going to investigate what affects how well a ball bounces. They should talk initially with a partner to decide what the variables are.

The children should be able to come up with a variety of ideas themselves, but prompt them if necessary, for example, would the surface the ball is dropped on or hit against make a difference? What about the material that the ball is made of, or from how high the ball is dropped? Explain that they will be working together in groups to plan their enquiry, identify which variables they will change or keep the same, and decide what they will measure or observe. They will then carry out their enquiry and report back on their findings at the end of the lesson.

Group children in three challenge groups according to how much expertise they have in planning and carrying out fair tests. A generic fair test planning grid is available as well as prompt questions (Resource sheet 1) to help children sequence the steps of their enquiry.

The challenges are presented on the Challenge slides to be displayed on the board, or printed out and placed in the centre of the table.

Challenge 1: Children work as a group to plan and carry out an enquiry to answer a question generated by the group

Provide the children with the Fair test planning grid (Resource sheet 1) to use as their group plans the enquiry.

Ask: *Which variables will you change and keep the same? What will you measure? How will you make sure your test is fair? Will you repeat the test? What evidence will you need to record?*

After the enquiry ask the children to share with the class what they found out.

Challenge 2: Children work as a group to plan and carry out their enquiry to answer a question generated by the group

Ask: *What is the question you are investigating? What are you planning to do to answer the question? Which variables will you change and keep the same? What will you measure? What evidence will you need to record? How will you record it?*

After the enquiry has been completed, prompt the children to evaluate their enquiry.

Ask: *Was your fair test enquiry successful? Did the method you came up with work? Did you collect sufficient evidence to help you answer your question?*

Challenge 3: Children work as a group to plan and carry out an enquiry to answer a question generated by the group

Ask: *What evidence will your enquiry need to provide in order to answer your question? How will you be sure your results are as accurate and reliable as possible? How will you record the evidence that you collect? What other sources of evidence might be useful and relevant to this enquiry?*

The children should think about further tests and additional research that might be relevant to answering their question. They may suggest repeating the test to confirm their results.

After the enquiry has been completed, prompt the children to evaluate their enquiry.

Ask: *Was your fair test enquiry successful? Did the method you came up with work? Did you collect sufficient evidence to help you answer your question? How could your enquiry be improved another time?*

REFLECT AND REVIEW:

Review the findings with children.

Ask: *What does your evidence show? How important do you think the material is in making a ball bounce well? Are there other things that need to be taken into account?*

Responses to these questions can then be used to lead into the next activity.

Children receive a letter (Resource sheet 2) from the manager of a new sports club in town. He is sorting out equipment for the club and wants to know the best balls to buy. Can children use what they know about balls, their materials and their properties to make some suggestions? What other variables will be important? Encourage children to think about the range of sports, budget, and number of people likely to be playing sports. Children can note their ideas and suggestions in a copy of the Sport4All table (Slideshow 1).

EVIDENCE OF LEARNING:

Listen to children's responses throughout the lesson and as they answer differentiated questions for each challenge group. In the Explore part of the lesson, how well did they contribute? Could they identify similarities and differences between balls? Were they able to identify their own criteria for sorting? For the challenges could they identify a question and describe the plan for a fair test enquiry developed by their group? Did they know which variable they were changing and which they were keeping the same? Did they make and record accurate measurements? Did they repeat readings? Could they relate their findings to the materials the balls were made from? Were they able to evaluate the success of their group's enquiry and suggest ways the outcomes might have been improved?

EVERYDAY MATERIALS

INTRODUCTION

In this module children further develop their knowledge and understanding of materials, achieving an in-depth knowledge of the properties of certain materials and how and why those specific properties make them suitable for particular uses. They explore familiar objects in detail and find out about accidental scientific discoveries, such as the 'non-sticky' glue developed by Spencer Silver and used in 'Post it' notes, and how properties of 'super absorbent powders' can make them useful in everyday life. Specific scientific and other vocabulary is used by children as they describe, explain and communicate their understanding of materials, succinctly and in ways appropriate to a science context.

When working scientifically, children plan and carry out comparative and fair tests to answer questions about how and why certain materials are selected and used because of their properties. They do this increasingly independently, recognising and controlling variables where necessary, so that they collect sufficient quality evidence to enable them to answer their science questions. Children take and record measurements using appropriate measuring equipment with increasing accuracy and use a variety of ways to report and present their data to an audience.

This module, together with Modules 3, 5 and 6, builds on earlier learning that began in Key Stage 1 and then continued in Year 4, Module 1, In a State. As part of the Year 4 module children compared and grouped materials according to whether they were solids, liquids or gases and learned about changes of state that take place when materials are heated or cooled. This series of modules offers the final chemistry-related learning for children in Key Stage 2. It provides teachers with ample opportunities to assess children's progress against the programmes of study.

Key vocabulary:

properties, material, building, construction, structure, organic, natural, manufactured, man-made, weathering, decay, decompose, break down, brittle, fragile, metal, plastic, wood, ceramic, concrete, compare, contrast, group, organise, criteria, strong, strength, weakness, durability, wear, tear, stretch, flexible, flexibility, hardness, light, heavy, durable, durability, waterproof, washable, stain resistant, reusable, bicycle, suspension, brakes, tyre tread, saddle, weight, mass, criteria, ovenproof, heat, temperature, room temperature, thermal conductor, thermal insulator, insulate, insulation, viscosity, viscous, sticky, stickiness, tackiness, adhesive, glue, saturated, powder, particle, polymer, volume, quantity

FACT FILE:

Materials that children encounter in the world around them show signs of wear and tear over time. This may be due to weathering or regular use (or abuse). Organic materials, for example wood, will decompose once the surface seal or varnish is broken, whereas some plastics start to break down and can become brittle.

Particular materials and their properties

Plastics: This term encompasses a wide variety of common everyday materials, all with very different properties. Plastics are polymers: very large molecules made up of smaller units joined together, generally end-to-end, to create a long chain. The properties of particular plastics depend on how these chains combine. Plastics can be very different: rigid and inflexible in one form, for example a plastic container; malleable and highly flexible in another, for example cling film.

Ceramic and glass materials: These tend to be strong, stiff, brittle, chemically inert and poor conductors of heat and electricity. Metal is a very good thermal conductor (as well as a very good electrical conductor), becoming hot quickly, but losing heat at a very fast rate, for example, cooling to room temperature much more quickly than ceramic or glass.

Glue: For two surfaces to become glued together, two things must happen. Firstly, the liquid glue must find its way into all the nooks and crannies of the two surfaces being stuck together. Even surfaces that feel quite smooth will appear much rougher when examined under a microscope. Once the glue has been applied to the surfaces it must be able to turn into a solid, either by a setting or a drying process. This solid then holds the two surfaces together. If it cannot dry or set for some reason, it will not stick.

Insta-Snow®: The scientific name for Insta-Snow® is sodium polyacrylate. It is a polymer that takes the form of a 'super absorbent powder'. It is very similar to the powder used to fill babies' nappies, which was originally developed by NASA and used by astronauts on space flights. Scientists developed a version of that polymer that was far too fluffy when it absorbed water to make it suitable for a nappy – and it became Insta-Snow®. The version of the polymer inside a nappy turns to gel instead – altogether more appropriate!

Insta-Snow® absorbs water by osmosis and the molecules in the chain swell – but only so far. As the powder absorbs water, an exothermic reaction takes place and the changing powder feels warm to the touch. If salt is added to the Insta-Snow® at this point the polymer 'gives up' its water and releases it, making the mixture much wetter.

The powder eventually becomes saturated when it has absorbed as much water as it possibly can. If the water is allowed to evaporate, the powder can be used again – although the impurities from tap water seem to make it less effective if reused.

Thermal properties of materials

A thermal insulator is a material that provides high resistance to heat flow, for example, types of foamed plastics like polystyrene, wood, some fabrics and cork. Polystyrene is a very good thermal insulator and is able to keep cool things cool and hot things hot. It has the added advantage of being resistant to moisture, mould and mildew. It is frequently used in cool bags.

Common misconceptions:

Children sometimes use the word 'material' to describe fabric and textiles. They need to be reminded that in science the word 'material' is a generic adjective used to describe what something is made of.

Many children believe that all metals are magnetic. Only metals containing iron (including steel), nickel and cobalt are magnetic (that is, can be attracted to a magnet).

EVERYDAY MATERIALS

 ## LESSON 1: WHICH MATERIALS ARE USED IN OUR SCHOOL BUILDINGS, WHAT FOR AND WHY?

Key vocabulary:

properties, material, building, construction, structure, organic, natural, manufactured, man-made, weathering, decay, decompose, break down, brittle, fragile, metal, durable, durability, plastic, wood, ceramic, concrete, insulate, insulation

LESSON SUMMARY:

In this lesson children identify a variety of materials in different forms, observing how they are used for specific purposes within school buildings. This lesson builds on learning in the Year 3 module, Rock Detectives, where children carried out a survey of how rocks are used around school, and in the Year 5 module, Get Sorted, where children grouped and classified different materials. By the end of this lesson they will have carried out a survey of building materials: identified how they are being used, noted any signs of wear and tear they observe, and suggested ways in which insulation and building maintenance might be improved.

Preparation required: Identify a suitable route around the inside and outside of school buildings.

National curriculum links:

Give reasons, based on evidence from comparative and fair tests, for specific uses of everyday materials, including metals, wood and plastic

Learning intention:

To recognise that materials are used in many different ways and for particular purposes within buildings

Scientific enquiry type:

Grouping and classifying

Working scientifically links:

Reporting and presenting findings from enquiries, including conclusions, causal relationships and explanations of and degree of trust in results, in oral and written forms such as displays and other presentations

Success criteria:

- I can identify the variety of different types of materials used around school.
- I can link the properties of the material to its use.
- I can give examples of where certain materials are showing signs of decay or wear and suggest why this might be.
- I can describe how buildings can be insulated and propose how insulation of school buildings might be improved.

EXPLORE:

Find out how much exposure children have had to building projects, either through personal experience or through the media.

Ask: *Have you any experience of a building project, perhaps a house extension, building in your locality, or from watching DIY TV programmes? What is done first? What types of materials are used for buildings? How are buildings constructed?*

Explain to children that it usually takes months to construct a building like a house or a school. Plans have to be drawn up, permission given for buildings to be built, a construction team put in place, and materials ordered and brought onto site before construction can start.

Show children the video of the different materials used in building construction (Video 1). Ask children to look out for the different construction materials that are being used. How are the materials put together? What are their different purposes? Show the video a second time, so that they can make notes.

ENQUIRE:

Explain to children that during this lesson they are going to carry out a survey of the materials that have been used in buildings around school, and review their current condition and how well they are insulated and maintained.

The challenges are presented on the Challenge slides to be displayed on the board, or printed out and placed in the centre of the table.

Challenge 1: Children identify different materials in use in buildings around the school

Provide the children with a Materials survey recording grid (Resource sheet 1) on which to record details of the materials that they identify during their survey. Explain to the children that they should record information about different types of materials and their location, and the reasons why the properties of those materials make them suitable for that particular use. On return to the classroom, give the children some time to organise and review their data.

Ask: *Which materials did you see? Which materials did you see most often? How were the materials being used? Were particular materials always used in the same way?*

Children should recognise the need for sub-groups within types of materials, for example, different types of plastic (rigid and soft), and reorganise their data to take this into account, with support if necessary.

Challenge 2: Children carry out a 'wear and tear' survey around school

Ask: *What do we mean by 'wear and tear'? What signs of wear and tear might you see?*

Ask the children to record information about different types of materials and their location around school, the amount of wear and tear they observe, and any ideas they might have about the causes of this wear, for example, exposure to extreme weather or regular use over time. The children should note their observations and ideas in a jotter or use the Materials survey recording grid (Resource sheet 1). On return to the classroom give the children some time to organise the data that they have collected, identifying where the most significant evidence of 'wear and tear' was seen.

Ask: *What do you think is the cause of the wear and tear you have observed? How could you find out more about the causes of damage to the materials around our school?*

Challenge 3: Children carry out tests to identify where insulation could be improved around school, document evidence of poor insulation and suggest materials that could be used to improve insulation

Explain to the children that it is important to have very good insulation in and around our schools. Not only does good insulation make everyone's time in school more comfortable, but it can also reduce the enormous cost of heating the buildings and help reduce energy wastage.

Ask: *Where do you think we lose most heat from our school buildings?*

The children should mention gaps around windows and doors, which lead to draughts.

Show them the Buildings insulation video (Video 2), which shows a simple 'tissue test' that can be used to identify where heat is lost and insulation may be poor. Children could carry out the 'tissue test' to identify or confirm where draughts exist in the classroom and around school.

Ask the children to record the evidence that they collect, producing either an annotated plan of the school building or a sequence of annotated photographs to show where insulation needs to be improved.

Ask: *What parts of the structure of our buildings have you found where insulation is poor? Do you know how insulation can be improved? Did the video give you any ideas? How might you use other secondary sources to help you find out more?*

Challenge the children to extend their enquiry and find out which materials might be used to improve the insulation of the school buildings.

Ask: *What types of materials are often used to insulate buildings? What particular properties do these materials have in common that makes them suitable to use?*

REFLECT AND REVIEW:

Encourage children to discuss in pairs what they found out about the school, and how the insulation and maintenance of the buildings might be improved. Share ideas across the class and develop a top 10 list of feedback points that should be included within the report to the head teacher.

Ask: *Do you need any further information or evidence from additional research? How could you present your evidence to the head teacher most effectively?*

EVIDENCE OF LEARNING:

Listen to children as they gather information and analyse the evidence that they have collected. Can they identify a wide variety of different materials used around school, describing how materials may vary in form, for example plastics of different types, and identifying why they are used for certain jobs? Can they give examples of where certain materials are showing signs of decay or wear and suggest why? Can they describe how buildings can be insulated and propose how to improve school insulation? How effectively do they identify key points to feed back within a report to the head teacher, as part of an Eco-schools Environmental Review?

Key information:

Materials used in building construction, although selected for use because of their durability, will show signs of wear and tear over time. This may be due to weathering or regular use (or abuse). Organic materials, for example wood, will decompose once the surface seal or varnish is broken, whereas plastics start to break down and can become brittle. Plastics are polymers – made up of chains of molecules joined together in long strings, which can begin to break down over time. Most metals, unless protected by paint or treated in another way, will start to oxidise, forming rust.

EVERYDAY MATERIALS

LESSON 2: WEIGHTY PROBLEM: WHICH IS THE BEST CARRIER BAG?

Key vocabulary:

properties, material, compare, contrast, strength, weakness, durability, wear, tear, stretch, flexibility, weight, mass, plastic

Resources:

Lengths of thick dowel, broom handles, etc., modelling clay, large masses, for example, bricks, heavy books or cans of food to test bags, stop watches, different types of carrier bags, thick and thin plastic

Health and safety:

Teach children how to lift items safely. Place a box containing soft or waste materials under any hanging load. The load itself should be as low as possible (see Be Safe! section 13).

LESSON SUMMARY:

In this lesson children plan and carry out a fair test investigation into different types of plastic carrier bags, building on from Module 5, Lesson 3 where they sorted, grouped and tested a wide range of plastics according to their properties. By the end of this lesson children will have tested the suitability of a variety of bags for different purposes and recommended the most suitable carrier bag for a specific purpose.

Preparation required: As a class, collect a wide variety of different plastic carrier bags in advance of the lesson, including some of the 'bags for life' produced by certain stores.

National curriculum links:

Give reasons, based on evidence from comparative and fair tests, for the particular uses of everyday materials, including metals, wood and plastic

Learning intention:

To plan a fair test to investigate different carrier bags and collect evidence to make recommendations regarding their use

Scientific enquiry type:

Carrying out comparative and fair tests

Working scientifically links:

Planning different types of science enquiries to answer questions, including recognising and controlling variables where necessary

Success criteria:

- I can plan a fair test, controlling variables as necessary, to investigate the best plastic carrier bag for a particular purpose.
- I can use evidence that I have collected to say why a particular plastic carrier bag is best for a specific task.

EXPLORE:

Ask: *Which is the best carrier bag to use to carry shopping?*

Explore with children what might be meant by 'best'. The strongest? Handles that don't mark your fingers? Both? Collect children's ideas on sticky notes and place on a poster with the heading: Variables to measure.

Provide children with a selection of plastic carrier bags to look at in detail. Ask them to look at the type of plastic, shape, size and construction, and identify any different features, for example, extra strengthening around the handle area or use of thicker plastic in 'bags for life'. Collect these ideas on sticky notes and place on a poster with the heading: Variables that can be changed. Encourage children to think about how carrier bags are used.

Ask: *Is shopping only being transferred to the car boot or is the shopping being carried a long way by hand? How thick are carrier bags generally? Are free carrier bags always thin and charged-for carriers always thicker? Does it depend on what the shopping is that goes inside?*

ENQUIRE:

Explain to children that three local shop owners have asked them to test a variety of carrier bags and to suggest which would be the best choice for their particular customers. Information about the needs of the customers of each shop is provided on Task cards (Resource sheet 1).

In all three challenges children make observations and collect evidence, and use this evidence to support recommendations they make to their clients. They should all refer back to the dependent and independent variables they listed on the posters in the Explore activity.

The challenges are presented on the Challenge slides to be displayed on the board, or printed out and placed in the centre of the table.

Challenge 1: Children are supported to plan and carry out a fair test

Ask: *What is Mr Bryant's problem? What do we know about Mr Bryant's customers? What would be the best carrier bag for them? Can you plan a fair test to find out how much the handles of different carrier bags will cut into a customer's fingers as they carry a heavy load of shopping?*

Support the children as they plan a fair test, using the generic Fair test planner. Explain to them that the variable they need to change is the handle.

Ask: *What differences are there between handles? What could you measure or observe? What do you need to keep the same?*

Suggest or model the following method if the children struggle to come up with a method of their own.

Stick thick modelling clay strips onto a thick length of dowel or broom handle. Hang carrier bags of shopping over the modelling clay and measure how deep the handles cut in after a period of time.

Ask the children to complete a Results table (Resource sheet 2) as they test four plastic carrier bags with different handles. The results can then be discussed.

Ask: *Which handle will be the most comfortable to carry? What properties of the material that the handle is made from make it comfortable? Are the handles of a particular design? Do thicker bags have more comfortable handles?*

Challenge 2: Children plan and carry out a fair test

Ask: *What is Miss Johnson's problem? What will you need to find out in order to help her? What is the property that you will be testing?*

The children need to plan a fair test to find out which plastic is best for making a carrier bag to hold a large mass. They should use the Fair test planner and record the variable that they need to change (what the plastic the bag is made from) and what they need to measure (distortion, breaking). If necessary, prompt the children to think about the variables that they are keeping the same as far as possible (size of carrier bag, type of handle).

The children should record their observations in a two-column table.

Ask: *Which material makes the strongest bag? Which carrier bag, if any, would you recommend to Miss Johnson? Is strongest necessarily best? Is there a danger if customers try to carry too much in one bag?*

Challenge 3: Children plan and carry out a fair test, gathering specific evidence

Explain to the children that they need to come up with a recommendation for a plastic carrier bag that meets Mr Hamid's specific requirements.

Ask: *What is Mr Hamid's problem? What is most important for Mr Hamid's customers? Is it the handles that are the problem or some other part of the bag? What type of plastic carrier bag would be best for his customers, and why?*

Challenge the children to come up with a question that they can use a fair test to answer, and then to plan and carry out the test independently. They will need to refer to the independent and dependent variables identified at the beginning of the lesson and identify control variables, so that they can record their method and findings accurately, to justify recommendations to Mr Hamid.

REFLECT AND REVIEW:

Provide children with the opportunity to feed back their findings and rehearse the recommendations that they will make to Mr Bryant, Miss Johnson and Mr Hamid.

Ask: *What property did you investigate to solve each client's problem? What evidence did you collect to help you make a decision about the most appropriate carrier bag to recommend? Which carrier bag did you find was the best for each shop to use?*

Encourage children to think more deeply about the evidence that they have gathered.

Ask: *Would you pass on any advice about how the carrier bags should be used by customers? Can they sometimes be overloaded? Can they be used more than once? What about the 'bags for life'? Did anyone test them?*

EVIDENCE OF LEARNING:

Do children make sufficiently detailed observations and describe the similarities and differences in detail, and identify particular design features? Can they describe the properties of the plastic used and how this affects its suitability for making a carrier bag? Can they plan a fair test increasingly independently and controlling variables as necessary, to investigate the most appropriate carrier bag for a particular purpose?

CROSS-CURRICULAR OPPORTUNITIES:

This lesson links to Mathematics, through weighing the shopping, and D&T through the evaluation of carrier bags' design features.

EVERYDAY MATERIALS

LESSON 3: WHICH IS THE BEST TYPE OF PLATE TO USE?

LESSON SUMMARY:

In this lesson children carry out a comparative test to investigate how the properties of materials that are used to make plates affect their suitability for use in different situations or contexts. By the end of the lesson they will be able to explain how the material that a plate is made from makes it better than another for a specific purpose.

Preparation required: Cone off some 'test areas' outside the classroom where children can carry out a plate drop test safely.

Key vocabulary:

properties, material, compare, contrast, group, organise, criteria, hardness, durability, waterproof, washable, stain resistant, reusable, ovenproof, heat, temperature, thermal conductor, manufacture

Resources:

A variety of plates made of different materials (as similar in size as possible), ceramic, glass, pyrex, metal, plastic (different types and thicknesses), paper/card (different types and thicknesses), wood, plates that children can test to destruction (no best china), tools for chip test, safety goggles, tomato ketchup or similar for stain test, electronic weighing scales

Health and safety:

Ensure that children wear safety goggles when dropping plates; take care when using equipment to scratch or chip plates and carry out tests on a tray or in a shallow cardboard box so that any chippings can be easily found and disposed of. For a drop test children will need to be supervised; they might test on different 'coned off' surfaces outside, for example, the grass, tarmac or pavement. The test area will need to be thoroughly cleaned afterwards. Refer to Be Safe! section 13, Testing things.

National curriculum links:

Give reasons, based on evidence from comparative and fair tests, for the particular uses of everyday materials, including metals, wood and plastic

Learning intention:

To plan and carry out comparative tests to find out which material is best for picnic plates

Scientific enquiry type:

Carrying out comparative and fair tests

Working scientifically links:

Planning different types of science enquiries to answer questions, including recognising and controlling variables where necessary

Success criteria:

- I can identify properties of materials used to make plates that make them suitable for different purposes.
- I can plan and carry out a comparative test to collect evidence to help me decide which material is most suitable for a plate for a specific purpose.

EXPLORE:

Show children Picnic poser (Slideshow 1), which shows a photograph of a picnic. The slideshow asks some questions about the properties that plates need for use on a picnic.

Ask: *What's happening? What's important about the plates you take to a picnic? What are they used for? How are they taken to the picnic? What properties do the plates need to have? Should they be reusable?*

Record children's answers on a flip chart.

ENQUIRE:

Explain to children that they are going to work in groups of three to investigate the properties of the materials used to make plates. Challenges 1 and 2 require children to test different properties for their suitability for picnic plates; the level of direction and support is greater in Challenge 1. Challenge 3 requires children to identify a property that will make a plate suitable for a specific purpose, and to plan and carry out a fair test to identify the best plate for the job.

The challenges are presented on the Challenge slides to be displayed on the board, or printed out and placed in the centre of the table.

Challenge 1: Children plan and carry out a series of simple comparative tests to find out which plates are the most suitable for a picnic

Ask: *What is it important to remember when choosing plates to take on a picnic?*

Remind the children of the earlier discussion. The plates need to be carried, but although paper plates are lightweight they can be wasteful and expensive. The children need to investigate three properties: weight, stain resistance and whether they can be washed and reused.

Help the children to plan a series of simple comparative tests to investigate these properties and to record their evidence in the Comparing plates table (Resource sheet 1).

Challenge 2: Children plan and carry out a series of simple comparative tests to find out which plates are made from the most durable material

Ask: *Why is it important that the plates used at a picnic are hard-wearing?*

Remind the children of the earlier discussion. Plates might easily be dropped or damaged as they are carried to and from the picnic site. Encourage the children to think of ways to test a range of plates made from different materials to find out how durable they are. The children need to discover what happens if plates are dropped, scratched or knocked against something else

Ask: *What test will you do? What equipment will you use? What evidence will you record? How will you do your test safely?*

The children should record their results in a table (if it would be helpful, provide the children with a blank table from the generic assets).

Challenge 3: Children identify a question that relates to the properties of plates made of different materials for use in specific contexts and then plan a series of comparative tests to find out which plate is most appropriate for the job

Give each group of three children a set of the Where are plates used? cards (Resource sheet 2) and ask them to pick at random from the selection of events or contexts, for example, a children's party, a camping trip, a mountaineering expedition, for hot food, for soup, at a wedding or to use in a microwave.

Ask: *What are the most important properties for plates used in this context?*

Explain to the children that they need to plan and carry out a series of comparative tests, collecting evidence to show which material makes the best plate for their context, and record their results in a table (if it would be helpful, provide the children with a blank table from the generic assets).

Ask: *What will you do? What equipment will you use? What evidence will you record? How will you carry out your tests safely?*

REFLECT AND REVIEW:

Encourage children to think about and organise what they now know about different plates made of different materials, their properties and how this makes them useful for different purposes.

Provide children with photographs of the plates that they have tested and some headings that they might think about as they organise their ideas. For example: 'durable – hard to scratch/smash', 'good thermal insulator', 'reusable/easy to clean'.

Ask: *Which plates would you group together under which heading? Are there any that fit into more than one group? What other properties of materials could be used to group the plates differently?*

Children might either work in threes with large sheets of paper to map their ideas or contribute to a class map that you record on the whiteboard.

EVIDENCE OF LEARNING:

Observe children as they plan and carry out their tests. Can they identify the properties of the materials used to make a selection of plates, explaining how these properties might make different plates suitable for different purposes? Can they plan and carry out comparative tests to decide which materials might be better to use for plates for a picnic or in another context? Are children aware of how to carry out the tests safely? Can they record their evidence in a table? Can they use their findings to explain which materials will make the best plate for a specific purpose?

Key information:

Ceramic and glass materials tend to be strong, stiff, brittle, chemically inert and poor conductors of heat and electricity. Metal is a very good thermal conductor (as well as a very good electrical conductor), becoming hot quickly, but also losing heat at a very fast rate, for example, cooling to room temperature much more quickly than ceramic or glass.

EVERYDAY MATERIALS

Key vocabulary:

properties, material, compare, contrast, criteria, heat, temperature, room temperature, thermal conductor, insulator, insulate, insulation, materials such as polystyrene, cork, shredded fabric, wood chippings

Resources:

Thermometers, data loggers with temperature probes, hot water or soup in plastic containers with lids that have holes to allow access of thermometer or probe, ice cubes or ice cream in similar sized boxes or containers, cooked hot jacket potatoes, cool bags to use for testing, plus a couple of cool bags for disassembling

Health and safety:

Take care when handling hot potatoes and hot soup or water. Follow guidance for heating foods and using thermometers (see Be Safe! section 8).

Key information:

A thermal insulator is a material that provides high resistance to heat flow.

A thermal conductor allows energy in the form of heat to be transferred within the material, without any movement of the material itself.

Materials that are poor thermal conductors can also be described as being good thermal insulators, for example, wood, plastic and cork.

LESSON 4: COOL BOX CONUNDRUM: CAN THE SAME CONTAINER KEEP COLD THINGS COLD AND HOT THINGS HOT?

LESSON SUMMARY:

In this lesson children investigate how a cool bag affects the temperature of hot and cold food. By the end of this lesson they will know how a cool bag acts as an insulator to slow down changes in temperature.

Preparation needed: Ahead of time, cook some jacket potatoes and keep them tightly packed in an insulated bag or box until the lesson starts, but do not let children see where you have kept them.

National curriculum links:

Give reasons, based on evidence from comparative and fair tests, for the particular uses of everyday materials, including metals, wood and plastic

Learning intention:

To use evidence from investigations to explain how a cool bag works as an insulator

Scientific enquiry type:

Carrying out comparative and fair tests

Working scientifically links:

Taking measurements, using a wide range of scientific equipment, with increasing accuracy and precision, and taking repeat readings when appropriate

Success criteria:

- I can use the line on a line graph to answer questions about temperature change.
- I can calculate a change in temperature over time.
- I can explain how insulation in a cool bag can help to keep hot things hot and cool things cool.

EXPLORE:

Explain to the class that some children are going on a picnic and the menu that they have decided on includes both ice cream and hot jacket potatoes! These children want to know whether the same type of cool bag could be used to keep the hot things hot as well as cold things cold.

Ask: *What do you think?*

Display Cool box conundrum (Slideshow 1). Ask children which of the statements they agree with and how they can test what they think.

ENQUIRE:

Explain to children that during this lesson they are going to find out how a cool bag affects both hot and cold food. It is likely that many children will think that cool bags only keep things cold or even cool them down further. In Challenges 1 and 2 children collect data that will confront these ideas. Children who already show a clear understanding of insulation at this stage should take on Challenge 3. The groups for these challenges can be up to six in number.

The challenges are presented on the Challenge slides to be displayed on the board, or printed out and placed in the centre of the table.

Challenge 1: Children use thermometers or data loggers with temperature probes to compare how the temperatures of two hot potatoes, one in a basic cool bag, and the other on a plate, change over 30 minutes

Ask: *How often will you need to measure the temperature? How quickly do you think the potatoes will cool? Which potato will cool faster? What difference do you think the cool bag will make to the temperature of the potato?*

While the children are measuring how the temperature of each potato changes, and logging the changes on the Challenge 1 results table (Resource sheet 1), ask them to draw and annotate a sketch to show what they think is happening to the heat in each potato.

Key information:

Make sure that children remember from Year 4 that temperature is a measurement of heat and is measured in degrees Centigrade. Check that they know how to use a thermometer and read the scale accurately.

Key information:

The thermometer should be positioned in the potato, so that it can be read by children, without opening the cool bag. If this is not possible, talk with children about the potential effect of repeatedly opening the bag.

Key information:

Although this isn't a fully controlled fair test, it is important that the volume of either hot soup or ice cream is the same, in order to make a clear comparison between the soup placed in a container inside the cool bag and the soup (in a similar container) acting as a control, outside the cool bag.

Challenge 2: Children use thermometers or data loggers with temperature probes to measure the changes in temperature of hot soup and frozen ice cream that are either placed in separate but identical cool bags or left on a table at room temperature

Hot water and ice cubes can be used instead to represent the soup and ice cream.

Ask the children to measure the temperature of the ice cream and soup every 5 minutes, for 30 minutes. They should decide who in their group is going to measure the ice cream and who is going to measure the soup. They should record their results on the Challenge 2 results table (Resource sheet 2).

Ask the children who have been recording the soup and ice cream temperatures to pair up and draw a line graph to present both sets of data. There will be four lines on the graph: one for each set of control data (room temperature soup and ice cream) and one for the data for each food kept in the cool bag. They can use the generic graph template.

Ask: *What do the line graphs show about the impact of the cool bag on the temperature of the hot soup?*

Ask the children to write a conclusion to explain whether their results show that a cool bag is better at keeping something cold or hot, or whether there is no difference.

Challenge 3: Children disassemble a cool bag and use what they find out to design a 'Super Cool Bag' of their own

Ask the children to carefully take a cool bag to pieces, and to look at the materials it is made from and the way it has been made. They should then use the information to design a 'Super Cool Bag'.

Ask: *What is the purpose of a cool bag? How are cool bags made and what are they made from? How does the insulation in a cool bag work? What materials are used for the insulation and what specific properties do those materials have that help them to keep hot things hot for longer and cool things cool for longer?*

Explain to the children that their particular challenge is to use what they know and have found out about insulation to design a bag in which both hot and cold food can be insulated at the same time well enough to maintain their temperatures. Provide them with large sheets of paper on which to draw and annotate their designs. Remind the children that their design needs to include details of the materials they use, the type of insulation they include and how this is arranged in the cool bag.

REFLECT AND REVIEW:

Ask children who have completed Challenge 3 to share the designs for their 'Super Cool Bags'. They should explain how particular design features they have included are intended to work.

Ask: *Can you explain how a cool bag can keep cold things cold and hot things hot? What do others in the class think? Does your evidence suggest that it should be possible? Did the cool bags you used in your tests work for hot and for cold things? What happened to the temperatures you recorded over time?*

Show Slideshow 1 again.

Ask: *Which statement do you agree with now? Have you changed your mind? Why?*

Give children time to talk with a partner and to come up with a conclusion that summarises what they found out. Ensure that their statements point out that the temperature of the hot foods went down (whether inside or outside the cool bags), while the temperature of the cold foods went up, but that the temperature changed more slowly and to a lesser degree for those items that were inside the cool bag.

Ask: *Is the name 'cool bag' the best name or can you think of a better one? Will the temperature change eventually stop?*

EVIDENCE OF LEARNING:

Observe children as they carry out their investigations, look at their tables, line graphs and designs, and listen to conclusions. Do they use a thermometer or data logger correctly? Can they record their results accurately and calculate the change in temperature over time? Can they plot their results on a line graph? Can they use their results to explain how the insulation material in a cool bag helps to keep hot things hot and cool things cool?

CROSS-CURRICULAR OPPORTUNITIES:

This lesson links to Mathematics, through calculating differences in temperature, and to D&T through designing and testing products.

EVERYDAY MATERIALS

LESSON 5: MYSTERY MATERIAL: WHAT WILL HAPPEN IF WE ADD WATER TO THE MATERIAL?

Key vocabulary:

properties, material, compare, contrast, absorb, absorbency, saturated, material, particle, polymer, volume, quantity

Resources:

Tub of 'Insta-Snow®' (available from TTS and other suppliers), water jugs, measuring cylinders, pipettes or water droppers, syringes, paper clips, jelly strings, hand lenses

Health and safety:

Ensure that children are aware that they must not taste the mystery material in order to try to identify it.

Key information:

The scientific name for Insta-Snow® is sodium polyacrylate. This chemical is a polymer that is very similar to the super absorbent material found inside babies' nappies, which was originally developed by NASA and used by astronauts on space flights. Scientists developed a version of the polymer that was far too fluffy when it absorbed water to make it suitable for a nappy – and it became Insta-Snow®. The version of the polymer inside a nappy turns to gel instead – altogether far more appropriate!

LESSON SUMMARY:

This is the first part of a two-part lesson. In this lesson children observe and measure the effects of adding increasing volumes of water to quantities of a mystery material. Some children use secondary sources of information to help them explain to others what they see happening. By the end of this lesson they will recognise the importance of measuring and recording observations carefully, and be able to explain how one type of polymer behaves when water is added.

Preparation required: This lesson is the first of a pair of lessons. For the second part of the lesson (Lesson 6), invite a parent with a baby to talk to children about nappies and how they are used. Try also to get hold of different brands of nappy packaging and plenty of samples of nappies for children to test.

National curriculum links:

Give reasons, based on evidence from comparative and fair tests, for the particular uses of everyday materials, including metals, wood and plastic

Learning intention:

To observe, measure, describe and explain the changes that happen to a mystery material when water is added

Scientific enquiry type:

Observing changes over different periods of time

Working scientifically links:

Taking measurements, using a range of scientific equipment, with increasing accuracy and precision, and taking repeat readings when appropriate

Success criteria:

• I can decide what observations I need to make and what measurements to take as I add water to the mystery material.

• I can select the best equipment for the task and use it accurately to measure the quantities involved.

• I can use scientific vocabulary to describe what happens to the mystery material as water is added. I can explain the processes involved when water is added to the mystery material.

EXPLORE:

Give each child less than a teaspoon of the mystery material in the palm of their hands and ask them to observe the material closely.

Ask: *What does the white material look like? Can you describe the particles? Does it remind you of anything similar?*

Ask children to add a small amount of water, a drop at a time, to the material. It is useful to film the changes and replay it later.

Ask: *What do you notice? What happens to the material? Can you feel anything? What if you move the material around? Does it absorb more water? How much water do you think a small amount of snow could absorb?*

ENQUIRE:

Explain to children that during this lesson and the next they are going to explore the properties of 'super absorbent materials'. In Challenges 1 and 2 in this lesson, children explore the volume of water a quantity of Insta-Snow® will absorb over time. In Challenge 1 they are supported to follow a simple test method and record their results in a table. In Challenge 2 children are expected to make increasingly accurate measurements and record the outcomes of their test more systematically. In Challenge 3 children communicate their understanding of what happens to Insta-Snow® as water is added, using diagrams and physical models to help them to share their ideas with others.

The challenges are presented on the Challenge slides to be displayed on the board, or printed out and placed in the centre of the table.

Key information:

The material eventually becomes saturated when it has absorbed as much water as it possibly can – around 30 times its own weight. If the water is allowed to evaporate, the material can be used again, although the impurities from tap water seem to make it less effective if reused.

Key information:

The word 'polymer' means a long chain of molecules (poly – many, mer – unit or molecule). Insta-Snow® absorbs water by osmosis and the molecules in the chain swell, but they do have a limit as to how far they can swell. As the material absorbs water, an exothermic reaction takes place and the changing material feels warm to the touch for a short time.

Key information:

If salt is added to Insta-Snow® when the polymer has absorbed water, it 'gives up' its water and releases it, making the mixture much wetter.

Challenge 1: Children make observations and record measurements as they investigate how much Insta-Snow® increases in volume when different quantities of water are added

Ask: *How could you find out how much our snow material increases in volume when different quantities of water are added? What equipment could you use? How much material should you use? What would you need to measure?*

Show the children selected pieces of equipment, for example, a syringe or a narrow measuring cylinder marked in millilitres. Help them to come up with a simple method for their test, for example, place a small quantity of Insta-Snow® in a measuring cylinder, add a measured amount of water using the syringe, and record how much the snow expands each time water is added (measuring the height of snow in the cylinder). Ask them to record their results in a table (Resource sheet 1). They need to decide on the quantities of water that they are going to add and record those on the table.

Challenge 2: Children identify what observations they need to make and what measurements they need to take as they find out how much water Insta-Snow® is capable of absorbing

Explain to the children that they are going to work in pairs to carry out a test to find out when Insta-Snow® has absorbed the maximum amount of water that it can. They need to record their results systematically so that they can compare what they find with the results of other children.

Ask: *How will you measure the amount of water you are adding and the volume of snow produced? How will you know when the snow has absorbed as much water as it possibly can?*

Challenge 3: Children research what a polymer is and create a model of a simple polymer using paper clips and jelly strings to help explain about polymers to the rest of the class

Provide the children with some basic information about polymers and explain that Insta-Snow® is made of polymers tangled up together. When water is added they swell up, a little like super-soggy strings of spaghetti. The polymers can absorb a large amount of water, but eventually they stop being able to absorb any more.

Provide the children with paper clips and jelly strings so that they can make their own polymer models.

Ask the children to work in pairs or in threes to find a way to explain to the rest of the class what happens to the polymers in Insta-Snow® when water is added. They should plan their explanation on paper and use diagrams and their models to help them make their explanation clear.

If the children reach the point where they have completed their explanation, introduce the idea that something happens to the water-saturated Insta-Snow® when salt is added. Can they find out what it is? What might be happening to those swollen polymers?

REFLECT AND REVIEW:

What is Insta-Snow® and how does it work? Children who completed Challenge 3 should share with the class what they have discovered about polymers and the process of change that takes place as polymers absorb water.

Ask: *How much water did your snow absorb before it couldn't take any more? How much snow was produced by adding different volumes of water? If a teaspoon contains 1 g of material, what volume of water might it be able to absorb? How much snow could be produced with a whole 100 g tub? The paper notes inside the tub suggest it will expand to approximately 100 times its original volume. What size tub would that need? As big as a dustbin?*

Children might explore some of these ideas further as a homework challenge.

Advise children that in the next lesson they are going to find out about one of the practical, everyday uses of a similar sort of polymer to that in Insta-Snow®, which is contained in babies' nappies.

Ask any children with babies in their families to talk to their parents about what they think is important in a good nappy and to bring the information to the next lesson.

EVIDENCE OF LEARNING:

Listen to and observe children's responses throughout the lesson. Can they decide what observations they needed to make and what measurements to take? Can they select or identify the best equipment to use for the task? Do they use it accurately? Can they describe what happened to the mystery material when water was added? Are they able to use appropriate vocabulary to explain their understanding of the processes taking place?

EVERYDAY MATERIALS

LESSON 6: NAPPY ENDING: WHAT'S THE BEST BRAND OF NAPPY?

Key vocabulary:

properties, material, compare, contrast, absorb, absorbency, saturated, powder, gel, polymer, volume, quantity, product, manufacturer

Resources:

Mini whiteboards, water jugs, measuring cylinders, pipettes or water droppers, syringes, a collection of nappies with a variety of brands

LESSON SUMMARY:

This is the second part of a two-part lesson. Children apply knowledge of the super-absorbent properties of materials that they learned about during Lesson 5. In order to fully investigate and product test the nappy brands, and to allow for fully developed feedback to parents, this lesson should be extended to a 2-hour slot.

In this lesson children investigate different brands of nappies, coming up with their own questions and methods of enquiry. They identify the evidence that they need to collect so that they can provide information to parents about the various brands of nappy on offer and the brand claims. By the end of this lesson children will have drawn conclusions based on their evidence and linked what they have discovered to their growing subject knowledge of materials and their properties. They will have communicated their evidence creatively within a presentation that they developed themselves.

National curriculum links:

Give reasons, based on evidence from comparative and fair tests, for the particular uses of everyday materials, including metals, wood and plastic

Learning intention:

To present findings from a comparative test of nappies as a recommendation for parents of babies

Scientific enquiry type:

Carrying out comparative and fair tests

Working scientifically links:

Identifying evidence that has been used to support or refute ideas or arguments

Success criteria:

- I can identify a question that I can investigate to test the effectiveness of nappies.
- I can plan how to collect evidence to answer my question about nappies.
- I can draw conclusions about the properties of the materials used and the effectiveness of the nappies tested.
- I can present a persuasive argument about why one brand of nappies might be better than another.

EXPLORE:

Invite a parent with a small baby to visit the class. Children prepare questions in advance of the parent's visit, using any information that they have gathered from home to help them identify what to ask. For example, what makes a good nappy? Why do you choose the brand that you buy? Are they all made in the same way or are they different for boys and girls? Do nappies come in different sizes? How much do they cost? How many do you use each day? How many each week? Can nappies be washed and used again? How do you dispose of them? Which brands are the best value?

Ask children to make notes on a mini whiteboard so that they can use the information later. Alternatively, you could show the video about nappies (Video 1), which answers the questions above.

ENQUIRE:

Show children the marketing information and claims on the packaging of different brands of nappies. If there are not enough sets of packaging for all to see easily, project an image of each in turn using a visualiser. Ask children to identify any information that might be useful, for example, does the packaging mention how much the nappy might absorb without leaking?

Explain to children to investigate different brands of nappies, test them to see whether they are effective and then present their evidence to a group of parents about which brand is best.

Ask: *What questions can you think of that might be useful to investigate?*

Given earlier discussions, ideas and questions shared, it is reasonable to expect at this stage that children will have no problem coming up with ideas. If they struggle, prompt them by showing them the Nappy question starting points (Slideshow 1), which features a selection of questions that children could use as starting points. Children decide which question to investigate. They plan their enquiry and think about what evidence they need to collect in order to answer their question. In Challenge 1 children are supported in choosing a question to investigate and in planning for their enquiry. In Challenge 2 children identify their own question and the best method to collect reliable evidence, in order to provide visiting parents with evidence-based recommendations. In Challenge 3 children plan an enquiry to test one of the claims featured on nappy packaging. They plan systematically, identifying the evidence that they need in order to support or refute the claim they have chosen to investigate, so that they can create a persuasive argument to present to the visiting parents.

The challenges are presented on the Challenge slides to be displayed on the board, or printed out and placed in the centre of the table.

Challenge 1: Children work in groups to plan a comparative test to answer a question of their own (or one from the list) about nappies and present their recommendations about the best nappy

Provide the children with the Fair test planning frame (Resource sheet 1) to help scaffold their thinking. They can use blank table and bar chart templates (see the generic resources).

After the investigation is complete support the children as they decide how to present their recommendation to parents of babies.

Challenge 2: Children in groups plan a comparative test to answer a question of their own about nappies

Ask: *What evidence do you need to collect from your practical investigation? How will you be able to use it? Can you gather other evidence from elsewhere that might help you make a case for the best nappy?*

After the investigation is complete, give the children time to collect and organise their results.

Ask: *What does your evidence show? Can you make a statement that summarises what you have learned about the nappies you have tested? How will you use your evidence to suggest which is the best nappy to a parent?*

Challenge 3: Children in groups plan a comparative test to investigate one of the claims on a nappy brand's packaging

Ask the children to identify a specific claim on one nappy package and then to come up with a plan to check if it has any basis. Encourage them to think about how they might test the claim.

Ask: *What might you investigate? What evidence would prove or disprove the claim? What kind of test/s should you do? What other evidence would be useful to help confirm the results of your practical test?*

Remind the children that they need to present their evidence to parents and make a strong case, based on the evidence they discover through their enquiries.

REFLECT AND REVIEW:

Identify a number of children, or groups of children, to feed back the outcomes of their comparative test investigations to a parent. Encourage them to think carefully about what evidence to present and give them a time limit for their presentation, to ensure that they keep focused. Children can be as creative as they wish, perhaps communicating their findings in a news bulletin or as a product advert, rather than as a test report. Children who completed Challenge 3 might present a 'consumer investigation report', styled along the lines of a TV documentary revealing the truth, or not, of claims made about a product.

EVIDENCE OF LEARNING:

Listen to children's responses, in particular during their final presentation of results. Have children identified any patterns in the data they collected? Have they summarised their results, making a general statement about what they found out? Have they used their scientific knowledge of materials and their properties in drawing conclusions? Have they presented a persuasive argument as to why one brand of nappies might be better than another? Were the arguments they made clear and supported by evidence? Did they refer to other evidence sources, beyond their practical test? Did they use scientific vocabulary effectively to describe or explain, when necessary?

EVERYDAY MATERIALS

 ## ENRICHMENT LESSON 1: ARE ALL BIKES THE SAME?

LESSON SUMMARY:

In this lesson children identify the variety of materials (and their properties) that are used in making bicycles of different kinds. By the end of the lesson they will have discovered how technology has improved the use of materials in bikes and led to the development of an extensive range of bikes, each specifically designed with a purpose in mind.

Preparation required: Warn children in advance that you are going to ask some of them to bring their bikes to school (if this doesn't normally happen). If you or a colleague have a bike of a different kind, for example, a mountain bike, track bike or collapsible bike, then bring that in too. Linking with a local bicycle store would be very useful, but would require you to give enough notice to the shop to ensure that someone can make the time to bring bikes to show children.

Key vocabulary:

properties, material, compare, contrast, criteria, hardness, light, heavy, durable, flexible, strong, suspension, brakes, tyre tread, saddle

Resources:

A number of bikes of different types, mini whiteboards, bike catalogues, bike advertisements

Health and safety:

Remind children to be careful when looking at and handling bikes.

Key information:

Even an inexpensive bike can be made of more than a hundred different materials. These include several kinds of steel, other metals such as chrome and aluminium, several kinds of rubber, a few oils, different types of plastic and carbon composites.

National curriculum links:

Give reasons, based on evidence from comparative and fair tests, for the particular uses of everyday materials, including metals, wood and plastic

Learning intention:

To identify a range of materials used in making bicycles and how the properties of those materials make them particularly suited to their use in different types of bikes

Scientific enquiry type:

Grouping and classifying

Working scientifically links:

Reporting and presenting findings from enquiries, including conclusions, causal relationships and explanations of and degree of trust in results, in oral and written forms such as displays and other presentations

Success criteria:

- I can identify different materials used in making different kinds of bikes.
- I can link the properties of these materials to their specific uses in different kinds of bikes.
- I can explain how technology in bike design has improved and the ways in which materials used in modern designs have changed and developed.

EXPLORE:

Bring a bike/s into the classroom. Children should look closely and make notes on mini whiteboards about what materials they think have been used to make the bike, how they have been used and why.

Ask: *What properties does a bike need to have?*

Encourage children to think about what a bike does and how it is used. At this stage they might identify more obvious answers, for example, needs to be strong, light in weight, hard-wearing, and have a good saddle and tyres.

Ask: *What materials have been used to make each part of the bike? What types of metals do you think have been used? How have they been used? How can we discover which parts of the bike might be aluminium? What might the seat be made of? What about the tyres?*

Use the KWHL grid (Resource sheet 1) to begin to summarise children's ideas.

ENQUIRE:

Explain to children that during this lesson they are going to discover much more about the materials that are used in bikes and how the properties of those materials make them suitable for the job they do.

Challenges are differentiated according to task and the degree of analysis and application of learning required. In Challenge 1 children produce an annotated diagram of a bike, detailing the variety of materials used and describing how the properties of those materials make them suited to the jobs that they do. In Challenge 2 children compare and contrast two different types of bike and identify in each case how materials are used because of their particular properties. In Challenge 3 children combine what they know already with any research they carry out, to design a 21st Century bike using high performance materials with specific properties suited to the bike's purpose.

The challenges are presented on the Challenge slides to be displayed on the board, or printed out and placed in the centre of the table.

Challenge 1: Children annotate a diagram of a bike, labelling the materials and identifying their properties

Provide the children with paper on which to produce annotated diagrams of one of the bikes. If any children find this too difficult, provide them with the Generic bike sketch (Resource sheet 2), as an outline sketch that they can adapt and label. Ask the children to label their diagram, showing the materials used in the different parts of the bike, and note why their properties make them suitable for the job they are doing. Encourage the children to add further relevant details where they can.

Ask: *What sorts of materials are used in your bicycle? What properties do those materials have that make them particularly useful for the job that they do?*

Focus in on a particular part of the bike. Encourage the children to think about what materials are used in the saddle, for example. What properties do those materials need to have? For the saddle, the outer layer needs to be waterproof and there needs to be some padding inside, for comfort.

Challenge 2: Children research different materials that might be used to build a bike for different purposes, for example, a child's bike, mountain bike, racing bike, track bike or collapsible bike

Ask the children to compare and contrast two bikes, using the Compare and contrast grid (Resource sheet 3) to help them record their ideas and the evidence they collect.

Ask: *How are the bikes similar? How are they different? What materials have been used in each bike? Are any parts made from completely different materials in each bike?* (For example, the bike might be metal, plastic or carbon fibre.) *Why do you think this might be? Have the different materials used in the bikes been selected specifically because of the job the bikes are designed to do?*

Challenge 3: Children use their knowledge and learning to design a 21st Century bike made from high performance materials, which has features designed for a particular purpose

Challenge the children to work individually or in pairs to use what they know already, combined with any research they might do, to produce a design for a 21st Century bike. Money is no object! Explain to them that they must decide what their bike is going to be used for and the technology that they need to include, and that they need to specify the high performance materials that would make the bike unbeatable.

Ask: *What high performance materials have you included in your design? What properties do those materials have that makes them particularly suitable for the type of bike you have designed?*

REFLECT AND REVIEW:

Play children the Bike history video extract (Video 1), showing how bicycles have changed over time.

Ask: *What materials were bicycles originally made of? How big were they? How heavy do you think they would have been? What are the most significant differences? How do the materials used to make bikes today make it possible for us to use them in different ways, travelling over lots of different surfaces, not just on paths and roads?*

Ask children to compare examples of bicycles from history with the sketches they have made of their visiting bikes and the new bike designs.

EVIDENCE OF LEARNING:

Listen to children's responses throughout the lesson. Can they make suggestions about the kinds of materials they think might be used in making bikes of different types? Does their knowledge increase as they research more about materials technology and how technology has changed the way bikes are made? Can children link the properties of materials to their specific uses in modern bikes? Can they explain how modern bikes have changed and developed, and compare them with the early designs they have learned about?

CROSS-CURRICULAR OPPORTUNITIES:

This lesson links to History, through studying the development of transport, and to D&T through evaluating the suitability of a product for a specific purpose.

EVERYDAY MATERIALS

ENRICHMENT LESSON 2: SPENCER SILVER AND STICKY NOTES: WHAT'S THE STICKIEST GLUE?

Key vocabulary:

properties, material, compare, contrast, criteria, stickiness, viscosity, sticky, adhesive, glue, tackiness

Resources:

Glue spreaders, different kinds of glue that are safe for children to use, for example, stick glue, PVA glue; paste, fabrics and other materials, for example, plastic, cellophane, card, felt, hessian, lolly sticks, sand paper, sticky notes and Post-it™ notes, 'surfaces', for example, sheets of plastic, carpet tiles, lino, ceramic tiles, felt or fabric squares; 'labels', pre-cut rectangles of plastic or fabric, milk, gelatine, flour, access to the internet or books for further research

Health and safety:

Remind children to wash their hands and avoid touching their eyes after handling glues.

Key information:

In 1968 a chemical scientist called Spencer Silver developed a reusable adhesive that didn't really stick. The glue he created could hold paper together, but wasn't strong enough to maintain the bond when pulled on. Unfortunately, the scientist was at that time trying to make a super-glue. It would take 12 years for him to use his non-sticky glue in the Post-it™ Note.

LESSON SUMMARY:

In this lesson children learn about the chemist Spencer Silver and how he created Post-it™ notes almost by accident, as he worked to create a super-sticky glue. By the end of this lesson children will have investigated and made glues themselves, identifying the properties of glue and how they can be used for different purposes.

The lesson may take longer than an hour as the glues used in the tests need time to dry in order to fully bond materials together. Perhaps an additional follow up 30-minute slot could be made available either later the same day or the next day.

National curriculum links:

Give reasons, based on evidence from comparative and fair tests, for the particular uses of everyday materials, including metals, wood and plastic

Learning intention:

To plan a test to measure the stickiness of different types of glue

Scientific enquiry type:

Carrying out comparative and fair tests

Working scientifically links:

Using test results to make predictions to set up further comparative and fair tests

Success criteria:

- I can describe how Spencer Silver used the results of his investigations to create a familiar product.
- I can use investigation evidence to measure different types of glues and their adhesive properties.
- I can use results from the investigations we have done to make predictions about further tests we could carry out.

EXPLORE:

The story of Spencer Silver.

Show children the video which tells the story of Spencer Silver (Video 1).

Ask: *Who was Spencer Silver? What was his job? What product was he developing when he made his accidental discovery? Why did it take him so long to find a use for his 'non-sticky' or reusable glue?*

ENQUIRE:

Explain to children that they are going to follow in the footsteps of Spencer Silver and investigate different kinds of glues, their stickiness and what they might be used for. In Challenge 1 children follow a method that is modelled to them first, while in Challenge 2 children have more flexibility to make choices, choosing from a wider variety of glues and materials to test. In Challenge 3 children make and test their own glue, using different proportions of key ingredients to see which results in the stickiest – and least sticky – glue.

The challenges are presented on the Challenge slides to be displayed on the board, or printed out and placed in the centre of the table.

Challenge 1: Children investigate different types of manufactured glues in order to find out which one is the stickiest

Provide the children with three different types of glue to test, glue spreaders and strips of material, i.e. plastic, cellophane, card, felt, hessian, cotton fabric and a lolly stick. Explain to the children that they are going to use two pieces of each material and see whether they stick to themselves (for example, test two strips of plastic and see whether they stick).

Ask: *How will we keep our test fair?*

Explain to the children that they are going to test three different glues, so they need to ensure that the same amount of each glue is used to stick all the samples of materials in the same way. They should use the Results table (Resource sheet 1) to record their results.

Challenge 2: Children decide on their own method of investigating different types of manufactured glues to find out which one is the best to use to stick a plastic or a fabric label onto different surfaces

Provide 'surfaces' for the children to experiment with, for example, sheets of plastic, carpet tiles, lino, ceramic tiles and felt or fabric squares. Provide ready-made 'labels' of plastic or fabric to save cutting out time.

Ask: *How will you set up your test? Which 'surfaces' will you use? Do you think the glues will all take the same time to stick? How long will you need to leave the glues before you check whether they have stuck or not? How will you record the evidence you collect? What will you need to record in order to answer your question?*

Challenge 3: Children make and test their own glue, using different proportions of key ingredients to see which results in the stickiest – and the least sticky – glue

Provide the children with milk, gelatine and flour, and a selection of materials to use to test their 'glues'. They need to mix together the milk, gelatine and flour in different proportions to create three different glues.

Ask: *Do you think a thick glue or a thin glue is likely to be stickier? What might be the problem if you create a very wet glue? How long do you think you will have to wait before you know whether the glue has stuck?*

Ask the children to mix their glues and set up their test. In order for the glues to stick, they need to return to their materials the next day.

In the meantime, ask the children to use secondary sources to find out about the stickiest glues on the planet. How might it be possible to stick a man to the ceiling? What could superglue be used for?

Key information:

For two surfaces to become glued together two things must happen. Firstly, the liquid glue must find its way into all the nooks and crannies of the two surfaces being stuck together. Even surfaces that feel quite smooth appear much rougher when examined under a microscope. Once the glue has been applied to the surfaces it must be able to turn into a solid, either by a setting or a drying process. This solid then holds the two surfaces together. If it cannot dry or set for some reason, it will not stick.

REFLECT AND REVIEW:

Give children who completed Challenges 1 and 2 the opportunity to feed back to the class their findings.

Ask: *What did you find out? Do all glues stick all materials? Do some glues work only on some materials? What does the packaging of the glues tell us? Why do some glues work better for some materials than others?*

Ask children who completed Challenge 3 to describe how they made their glues and explain which ones worked best on which materials.

Ask: *Did your glues work? Which worked the best? Which materials could you stick? What other tests could you carry out to find out whether there are other – and stickier – glues around? How might you test them? Why would it be useful to know?*

Show children the Sticky glue video (Video 1).

EVIDENCE OF LEARNING:

Listen to children as they investigate the properties of glues. Can they describe how Spencer Silver used the results of his investigations to create a familiar product? Can they link his work in developing glues to the glue products they have investigated/made and how they work? Can they use the evidence they collect during their investigations to compare different types of glue? And to describe their adhesive properties? Can children use their results to make predictions about further tests they could carry out and suggest alternative uses for glues of different types? Can they explain why we need different types of glues in terms of the properties of the different glues and their specific uses as a product?

CROSS-CURRICULAR OPPORTUNITIES:

This lesson links to English through the biography of the glue scientist Spencer Silver, and to Art through the creation of mini collage artwork.

MARVELLOUS MIXTURES

INTRODUCTION

In this module children further develop their conceptual knowledge and understanding of how different mixtures of solids and liquids might be separated. They learn that certain solids dissolve while others do not, and how these dissolved solids might be retrieved from a mixture. They explore how the rate at which solids dissolve can vary, investigating variables that might make a difference. They use their knowledge of separating mixtures in solving a number of real world-based enquiries, which require them to apply their growing subject knowledge to an unusual context. Children use specific scientific and other vocabulary as they describe, explain and communicate their understanding of materials, succinctly and in ways appropriate to a science context.

When working scientifically, children plan different types of enquiries to answer questions, recognising and controlling variables where necessary. They will use a range of science equipment with increasing accuracy and precision, and use a variety of ways to report and present their findings to an audience.

This module, together with Modules 3, 4 and 6, builds on earlier learning that was begun in Key Stage 1 and then continued in Year 4. There, children compared and grouped materials according to whether they were solids, liquids or gases and learned about changes of state that take place when materials are heated or cooled. This series of modules offers the final Chemistry-related learning for children in Key Stage 2. It provides teachers with ample opportunities to assess children's progress against the programmes of study and in readiness for end of key stage assessment.

Key vocabulary:

material, compare, contrast, separate, mixture, sieve, filter, evaporate, solid, liquid, gas, powder, particle, dissolve, soluble, solution, contamination, contaminate, contaminated, impurity, pure, purity, suspension, saturated, saturation, reversible, non-reversible, microbes, bacteria, types of oil, liquid, solid, detergent, sticky, filter, mechanical, boom, residue, environment, biological, marine life, purify, drinkable, sterilise

FACT FILE:

Many breakfast **cereals** are fortified with iron particles as a mineral supplement. Iron is found in a very important component of blood called haemoglobin. Haemoglobin is the compound in red blood cells that carries oxygen from the lungs around the body. It is the iron in haemoglobin that gives blood its red appearance. All the iron in a single human body would be enough to make two small nails!

Salt (chemical name – sodium chloride) gives us chlorine for the chlor-alkali chemical industry – the biggest user of salt. Chlorine-based chemistry provides clean water, soaps and detergents, many medicines, PVC pipes for homes, mobile phones, cosmetics, protective suits for divers and astronauts, digital cameras, flat panel TVs, electron microscopes and solar panels for energy production. The list is essentially endless.

Dissolving takes place when two materials, a solid and liquid, share a similar chemical property. **Salt** and **water** molecules carry positive and negative charges. As salt is mixed into water, the charged water molecules break apart the charged salt molecules. The sodium and chloride molecules mix uniformly with the surrounding water molecules and the salt dissolves in water, forming a mixture.

Oil, on the other hand, is made up of chains of molecules that are not charged – so are very different from both salt and water molecules. **Salt** will not dissolve in oil. **Sugar**, however, dissolves very slowly into the oil (this might be hard for the children to see). Its physical properties allow it to 'hide' within the oil molecules.

If sugar or salt is mixed with water, both solids dissolve because they are water-**soluble**. In a **solution** the particles of sugar or salt are evenly spread and light passes through the mixture, which is **translucent**. When solids do not dissolve, their particles cloud the mixture and form a **suspension**, such as flour in water, and the suspended particles can be removed using a filter.

There is a limit to how much of a solid can be dissolved in a given amount of water. When no more of the **solute** (salt, sugar) can be dissolved in the **solvent** (water) the solution is said to be saturated. If the water is heated, more salt or sugar will dissolve. As the liquid cools some solid will reappear from the solution (such as the sugar at the bottom of a cold mug of tea).

Sugar and salt both dissolve in vinegar; sugar dissolves faster than salt.

More solid will dissolve in water as the temperature of the water increases.

The rate of evaporation will speed up if there is an increase in the temperature of the liquid or air around that liquid, the surface area of the liquid, or the movement of the air over the liquid. The water from a solution will turn to water vapour as its temperature increases. The water vapour will condense on a surface that is cooler than the vapour.

Common misconceptions:

Children will often use the word 'disappear' interchangeably with dissolve, because they cannot see the solid once it has dissolved. Using a coloured sugar and showing them what happens to the water's colour, as the sugar dissolves, may help. Evaporating the liquid and retrieving the solid will demonstrate that the solid is still present and has not 'disappeared'. Some children may use the word 'melt' instead of dissolve. The confusion here may result from placing ice cubes in water. Point out that if the solid has 'melted', evaporating the liquid to retrieve the solid will not work. Try it!

MARVELLOUS MIXTURES

LESSON 1: HOW CAN WE SEPARATE MIXTURES?

LESSON SUMMARY:

In this lesson children are introduced to the idea that materials can mix in different ways and that they can be separated. They make their own sieves to separate a complex mixture of dry solids. By the end of this lesson they will have separated a range of solids from the Cupboard Catastrophe mixture, using their own sieves and procedures to do so.

Key vocabulary:

material, compare, contrast, separate, mixture, sieve, filter, evaporate, solid, liquid, powder, particle

Resources:

Disposable plates of different kinds – these can be pierced with nails, hole punches or bodkins to form makeshift sieves; selection of fabrics, nets and gauzes; Cupboard Catastrophe mixture – rice, raisins, large pasta, flour, dried lentils, dried peas, fine sand, white sugar, paperclips, wood shavings, plus three or four plastic spiders; large foil trays, plastic beakers, magnets, spoons

Key information:

When different materials come into contact they sometimes react to form other substances (see Module 6) but often they do not react and so remain unchanged. These mixtures can be separated using different techniques that make use of the properties of the materials.

Key information:

The way in which the materials can be separated depends on the properties of the materials: sieving materials of different sizes, filtering solids from liquids, use of a magnet, evaporation for substances that dissolve (see Lesson 4).

National curriculum links:

Use knowledge of solids, liquids and gases to decide how mixtures might be separated, including through filtering, sieving and evaporating

Learning intention:

To explain that materials can mix and to demonstrate that mixtures of solid materials can be separated by the technique of sieving

Scientific enquiry type:

Grouping and classifying

Working scientifically links:

Planning different types of scientific enquiries to answer questions, including recognising and controlling variables where necessary

Success criteria:

- I can explain that materials can be mixed but often they can be separated.
- I can describe the process of sieving mixtures to remove particles of different sizes.
- I can successfully separate a complex dry mixture, identifying and separating the materials or explaining why they are impossible to separate in this way.

EXPLORE:

Set up a small demonstration of pairs of materials, such as rice and peas, paperclips and small plastic cubes, sugar and sand, gravel and water, salt and water.

Ask: *What do you think these pairs of materials are? What would happen if we mixed them together? Do you think it is possible to separate them again? How might this be done?*

Establish that the combinations can be separated in different ways. Explain to children that in this lesson they will be looking at the dry mixtures. Those involving water will be looked at again in the next three lessons.

Some of these and other materials might be explored using Interactive 1, in which children can match mixtures to equipment to see if their suggestions would work, but it is essential that they have first-hand experience of the mixtures above first.

ENQUIRE:

Set the scene for the challenges by presenting children with the debris from a 'Cupboard Catastrophe'.

Explain that a shelf has collapsed in the school storeroom (or science cupboard) and several of the containers (between 4 and 10) full of dried materials have been tipped out and mixed together. The caretaker, intending to help, has made things worse by sweeping up all the remains, including the odd dead spider, and putting them into this box.

Ask: *Can you use your science knowledge to help sort out the mess?*

Explain to children that they can use the science equipment they have around the classroom, if it will help them to separate the mixture, but children from another class are using the sieves today, so they will need to make their own.

The challenges are differentiated by the complexity of the sieving task and the support children receive for the report provided. Make sure that children who show a clear awareness of how to separate the materials in the Explore activity are challenged to work out a way to separate a more complex mixture.

The challenges are presented on the Challenge slides to be displayed on the board, or printed out and placed in the centre of the table.

Challenge 1: Children separate solids from a complex dry mixture using sieves that they make themselves

Prompt the children to think about the order for separating each Cupboard Catastrophe solid from the mixture.

Ask: *Are there any materials you can remove easily before you begin sieving?*

They may suggest removing the large pasta and the spiders by hand, and removing the paperclips using a magnet. Give the children a paper plate to adapt to act as a sieve. Show them quickly how they could puncture it using a variety of tools to allow solids of different sizes to pass through, but encourage them to come up with their own ideas. Provide the children with the Sort recording grid (Resource sheet 1) to use to list each solid they manage to separate from the mixture, and ask them to place each collected solid in separate containers.

Challenge 2: Children separate the solids from the Cupboard Catastrophe mixture, in as few steps as possible

Encourage the children to think of ways to improve the separation of the materials.

Ask: *Which solids will you remove first? How? Do you need more than one sieve? How might you adapt your paper plate to make an effective set of sieves? How will your sieves differ? Are there any other materials you could use to make a sieve?*

If necessary, prompt them to look at fabrics of different kinds that might work.

Ask: *Are there any solids that you will find difficult to separate with a sieve? What other ways might you use to separate them out?*

Ask the children to draw a flow chart to show the order in which they separated the different materials and the equipment they used.

Challenge 3: Children separate the solids from the Cupboard Catastrophe mixture to get the cleanest separation possible – a 'super sieve' system

Ask the children to design a system of sieving using a large foil tray that can separate all the different solids, with their varying particle sizes, from the mixture. They may use additional resources, such as a magnet and some stiff card, to separate out sections of the tray to complete their challenge. When they have created their super sieve system, they should place each solid in a separate container.

Ask: *How will you adapt your sieves? What will you separate first? How does changing the hole size as you go through the separation process help? What properties of the materials have you used to help you separate them?*

Ask children to draw a labelled diagram of the super sieve showing how it is made and what property of the material allows it to be separated by a part of the super sieve.

REFLECT AND REVIEW:

Bring children back together to discuss their findings.

Ask: *How did you separate the ingredients? How successful was your method? Were all the solids clearly separated? Did you manage to retrieve them all? Were there any solids that you could not retrieve using a sieve?*

Through this discussion, the children should realise that particles that pass through sieves are smaller than the holes in the sieve itself. Therefore, sieves can be designed to let specific sizes of particles pass through.

Show children an animation of the process of milling flour (Animation 1). The miller sieves the flour repeatedly during the process. He says that at each stage of the milling process, flour is removed and bagged up, ready for use in bread making.

Ask: *Why does the unprocessed flour have to be sieved more than once? What happens to the material that does not go through the sieve? Why is it important for the flour to be as fine as possible?*

EVIDENCE OF LEARNING:

Listen to children's responses as they explore ways of separating the mixture. Can they describe the process of sieving mixtures to remove particles of different sizes? Can they explain that the smaller the size of the solid they need to separate, the smaller the holes in the sieve have to be? Are they able to successfully separate the Cupboard Catastrophe mixture, identifying and separating all 10 materials? Do they recognise that some materials, such as sugar, are impossible to entirely separate by sieving? Can they suggest a way that the sugar could be separated from the mixture? Do they recognise that materials can be separated by using differences in their properties?

MARVELLOUS MIXTURES

LESSON 2: WHAT HAPPENS WHEN WE MIX LIQUIDS AND SOLIDS?

LESSON SUMMARY:

This is the first of three lessons exploring the mixing and separating of liquids and solids. In this lesson children investigate dissolving solids. By the end of this lesson they will have tested a variety of solids and identified those that dissolve and those that do not. Some will have also investigated a variety of solvents or the saturation point of dissolved solids in water.

Key vocabulary:

material, mixture, compare, contrast, separate, sieve, filter, evaporate, solid, liquid, gas, powder, particle, dissolve, soluble, solution, solute, suspension, saturated, reversible, non-reversible

Resources:

Sand, salt, fruit syrup, brown sugar, large transparent beakers, collection of solids – powder paint, flour, sugar, sand, coffee granules, bath salts, tea leaves, baby powder, sugar substitute, bicarbonate of soda; collection of solvents – oil, vinegar, water; beakers, spoons, weighing equipment, measuring jugs

Key information:

The sand eventually settles to the bottom of the container but the salt dissolves. Children often use the word 'disappear' interchangeably with dissolve, because they cannot see the solid once it has dissolved.

Key information:

The levels of water will rise when sand and fruit syrup are added to the respective beakers. When the sugar is added there may be a rise in level when it is first added, but once it has dissolved there will be little change (if any).

National curriculum links:

Know that some materials will dissolve in liquid to form a solution, and describe how to recover a substance from a solution

Learning intention:

To identify through investigation some solids that dissolve and others that do not, and describe how to tell that a solid has dissolved

Scientific enquiry type:

Grouping and classifying

Working scientifically links:

Using test results to make predictions to set up further comparative and fair tests

Success criteria:

- I can identify and name some solids that dissolve and some that do not.
- I can describe my observations accurately using key vocabulary, such as using the term 'dissolve' correctly, distinguishing between a solid that forms a suspension and one that dissolves.
- I can use what I have found out to make predictions for further tests.
- I can identify when a solution has become saturated and explain why.

EXPLORE:

Show children the pairs of materials that include water from the last lesson: sand and water, salt and water and ask children to recall what happened when they mixed the materials.

Encourage children to watch carefully, then in turn mix the water with sand and then the water with salt. Can children explain what is happening? Ask them to compare what was the same and what was different for the sand and salt, and whether they think they can get the solids back from the water.

This can be easily demonstrated for sand by filtering, which is the same as sieving used in the previous lesson. Explain that they need to think a little more about what happens to the salt before they can try to separate it.

Ask: *Can you think of any other solids that 'disappear'/dissolve when they are put into water? Why do you think this happens?*

Show children a number of beakers containing water. Mark the level of the water. Ask them to predict what will happen to the level of the water if you add sand to one of the beakers.

Ask: *What will happen if I add fruit syrup? What about some brown sugar?*

Ask them to draw quickly on a whiteboard or sketchpad what they predict will happen and share their responses. These will give you an idea of the level of understanding of dissolving across the class. Now add similar amounts of the materials in turn to separate beakers and mark the new water level. Are they surprised? Can they explain what they have seen? Gather ideas at this stage and return to them at the end of the lesson.

ENQUIRE:

Show the class a selection of solids to investigate during this lesson. Each group of children working on separate challenges can then contribute to a discussion to compare findings at the end of the lesson.

Provide individual challenge groups with a manageable selection of solids and liquid solvents to test. The challenges are differentiated by the complexity of the task. Encourage the children to take a challenge that extends the understanding they showed in the Explore activity.

The challenges are presented on the Challenge slides to be displayed on the board, or printed out and placed in the centre of the table.

Challenge 1: Children test a variety of solids to find out whether they dissolve in water, and use key vocabulary to describe and explain their observations

Show the children a selection of solids (see suggestions in resource list) that they might test to discover which dissolve. Ask them to place a small amount of each solid in a transparent beaker or plastic cup and watch carefully as they add a small amount of water (initially, without stirring). Provide them with Do solids dissolve? (Resource sheet 1), one between two, on which to record their observations.

Ask: *What did you notice when you added the water to the solid? Were there any changes straight away? Did stirring your solid help? Which dissolved? How do we know it has dissolved? What happened to those that did not dissolve?*

Challenge 2: Children test a variety of solids (brown sugar, salt, sand and flour) to find out whether the liquid makes a difference to whether the solid dissolves, and use key vocabulary to describe and explain their observations

Provide the children with a series of different liquids to test in addition to water, such as oil and vinegar. If necessary, suggest a simple method (see Challenge 1); otherwise allow them to set up a test themselves.

Ask: *How will you set up your test so that you can compare your results for each liquid? You will need to record your observations in detail; how will you do this? What evidence will you need to collect? How will you know whether each solid has dissolved?*

Allow the children to complete their test.

Ask: *Were there any surprises in your results? Did the solids that dissolve in water dissolve in other liquids too? What if you warmed up the liquids? Would it make a difference?*

Challenge 3: Children investigate how much of a solid can dissolve in a measured amount of water, record their results in a table and produce a series of diagrams to explain their findings

Ask: *How much of a solid can dissolve in a liquid? What would you estimate? Is it the same for each solid?*

Introduce the idea that when no more solid can be dissolved into a liquid, the solution is said to be 'saturated'.

The children should compare both salt and sugar, working with no more than 50 ml water and solids to begin with. They should make detailed observations and record the quantities of solid they add systematically.

Ask: *How will you know when the solution is saturated? What will you see? What will the solution feel like? Do you think that more or less of the solid would dissolve if the water were hotter? Why?*

When they have completed their test they should draw and label diagrams to explain what has happened.

Key information:

Solutions are clear and transparent but may be coloured by the dissolved substance. Insoluble substances either settle to the bottom of the container or those made of small particles may remain floating in the liquid forming what is called a suspension, thus making the liquid cloudy.

Key information:

Salt and sugar dissolve in water. Salt will not dissolve in oil; sugar dissolves very slowly and might be hard to see. Sugar and salt both dissolve in vinegar but the sugar dissolves faster than salt.

REFLECT AND REVIEW:

Play True/False/Maybe so (Interactive 1).

Display the interactive and ask children to suggest which statement in turn should be placed into which category, justifying their choice each time. Some of the statements cannot be answered directly from the work done in this lesson – note children's answers and reasons and explain that they will investigate these in the next lessons.

EVIDENCE OF LEARNING:

Listen to children's responses throughout the lesson, particularly during the Reflect and review part of the lesson. Can they identify and name some solids that dissolve and some that do not? Do they record their observations accurately? Do they describe their observations accurately and use key vocabulary appropriately, such as using the term 'dissolve' correctly? Do they recognise that some solids will dissolve in some liquids but not others? Can they identify when a solution has become saturated and explain why? Can they suggest ideas for further investigations?

MARVELLOUS MIXTURES

 ## LESSON 3: WHAT MAKES A DIFFERENCE TO HOW FAST SUGAR OR SALT DISSOLVES?

LESSON SUMMARY:

This is the second of three lessons exploring the mixing and separating of liquids and solids. In this lesson children investigate what makes a difference to how rapidly a solid dissolves. By the end of this lesson they will have planned and carried out a comparative test in response to a challenge question, identified variables and collected evidence systematically in order to answer their question. They will also have described how to retrieve a dissolved solid from a solution.

Key vocabulary:

material, compare, contrast, separate, mixture, sieve, filter, evaporate, solid, liquid, gas, powder, particle, dissolve, soluble, solution, suspension, reversible, non-reversible, variable

Resources:

Rock salt, table salt, icing sugar, Demerara sugar, granulated sugar, water, disposable transparent beakers, saucers, teaspoons, measuring equipment, timers, hand lenses, mini microscopes

National curriculum links:

Know that some materials will dissolve in liquid to form a solution, and describe how to recover a substance from a solution

Learning intention:

To identify, through investigation, some variables that affect the rate at which salt or sugar dissolves

Scientific enquiry type:

Planning comparative and fair tests

Working scientifically links:

Planning different types of scientific enquiries to answer questions, including recognising and controlling variables, where necessary

Success criteria:

• I can identify variables that might affect the rate at which a solid dissolves.
• I can predict which variable I think will make the most difference.
• I can plan a comparative to investigate a question about the dissolving rate of salt and/or sugar.

Key information:

The time taken for a solid to dissolve depends on several variables including: the size and shape of the pieces (small granular forms dissolve faster than in 'lumps'); the amount of stirring; and the temperature of the liquid (usually an increase in temperature increases the rate).

EXPLORE:

Use the drag and drop interactive (Lesson 2, Interactive 1) from the previous lesson to remind children of the discussion they had at the end of Lesson 2 and the true/false statements they discussed.

Ask them to share their ideas about what will make a difference to the time it takes for a solid to dissolve, and prompt with further questions, if necessary:

Ask: *What will happen if we change the temperature of the water to dissolve a solid? Will the difference be large or small? What difference will stirring make? Would the amount of solid make a difference? The volume of liquid we use? The shape and size of the pieces of the solid?*

At this point, encourage children who investigated saturation point during Lesson 2 (Challenge 3) to explain that there is a limit to how much solid will dissolve in a given volume of water.

ENQUIRE:

Explain to children that during this lesson they are going to work together to investigate some variables that affect how quickly a solid dissolves in a liquid.

In all three challenges children begin with a broad starting question and determine which aspect of that question they will investigate, either as a group, or working in pairs.

The challenges are presented on the Challenge slides to be displayed on the board, or printed out and placed in the centre of the table.

Challenge 1: Children will set up a comparative test to explore an aspect of the question: Does the type of sugar make a difference?

Provide the children with a variety of sugars to observe closely using hand lenses or mini microscopes. Ensure that they recognise that the sugars have particles of very different sizes.

Ask: *Will all the sugars dissolve? Will any dissolve quicker or slower than the others?*

Ask the children to predict which and make a note of it on their Does the type of sugar make a difference? grid (Resource sheet 1).

Explain to the children that they need to come up with a test to compare the time taken for different sugars to dissolve in water at room temperature. If necessary, prompt them with a simple method: Use clear disposable beakers, the same amount of water and 1 teaspoon of each type of sugar in each beaker.

Ask: *What would be the best equipment to use to measure the time it takes for the sugar to dissolve?*

Allow the children to complete their test.

Ask: *Did all the sugars dissolve? Which dissolved quickest? Which slowest? Why do you think that might be? How could you speed up the process?*

Challenge 2: Children set up a comparative test to explore an aspect of the question: What makes the most difference to the rate at which salt dissolves?

Explain to the children that they need to come up with a test to find out what makes the most difference to the time taken for salt to dissolve in water. Remind them about the variables identified earlier.

Ask: *What do you think might make the most difference? Why? How will you test your prediction? What equipment will you use?*

Prompt them, if necessary, to test several methods (separately) while keeping everything else the same, for example, using the same quantity of salt and volume of water, changing the temperature of the water or the number of stirs or the type of salt. Allow the children to complete their tests.

Ask: *Which variable made the salt dissolve fastest? Which had the least effect? Why do you think that might be?*

Challenge 3: Children set up a comparative test to explore an aspect of the question: Which dissolves at a faster rate: salt or sugar?

Provide the children with a selection of salt and sugar (of different types and crystal size). Prompt them by asking questions as they plan.

Ask: *What do you think will affect the time taken for the solid to dissolve?*

The children will probably repeat points made earlier: stirring, the temperature of liquid and the amount of solid. Explain that they are comparing sugar and salt: what other factor might make a difference? Encourage them to think about the particle size of each type of solid: comparing rock salt and caster sugar might distort their evidence. Why? The children should plan their enquiry and systematically record their results, including numerical data. Allow the children to complete their test.

Ask: *What is the difference in the time the two solids took to dissolve? Are your results reliable? Could anything have affected them? How could you find out more, such as about how each solid dissolves (at a microscopic level)?*

REFLECT AND REVIEW:

Review the true/false statements on the drag and drop (Lesson 2, Interactive 1) in order to consolidate understanding of which variables affect how quickly something dissolves. Do children change any statement's position? What reason do they give?

In preparation for Lesson 4, show children a slideshow of cartoon characters discussing their dissolving solids investigation (Slideshow 1).

Ask: *How can we retrieve a dissolved solid? Do you think that, once a solid is dissolved, you can't get it back? If you filter the solution, do you think you can get some back? If you let the water evaporate, will the solid be left behind? Will it look exactly the same as it did before it dissolved?*

Ask children for feedback. Their responses and explanations will provide useful information about their understanding of how materials can be separated.

Ask children to pour some of their dissolved salt and sugar solutions from their investigations into shallow containers (saucers are ideal), place them in different places around the classroom, and observe and record the changes over the next few days in preparation for the next lesson.

Ask them to think about what difference the volume of solution in the saucer will make to the time it takes for the solid to reappear. Which variables could make a difference to how quickly they see the solid? Children might raise several variables: the amount of solid dissolved in the solution (its concentration); the temperature of the room; and the position of the saucer in the classroom.

EVIDENCE OF LEARNING:

Listen to children as they plan and complete their enquiries and to their responses during the Reflect and review part of the lesson. Could they identify variables that affect the rate at which a given solid dissolves at the beginning of the lesson? Were they able to predict which variable they thought would make the most difference and suggest why that might be the case? Were they able to contribute to the planning of a comparative or fair test to investigate the group's question?

MARVELLOUS MIXTURES

Key vocabulary:

material, compare, contrast, separate, mixture, sieve, filter, evaporate, solid, liquid, gas, powder, particle, dissolve, soluble, solution, suspension, reversible, non-reversible, contamination, microbes, bacteria

Resources:

Large bowls, saucers, salt solution, water jugs, desk lamps or other strong light sources, cling film, plastic sheeting

LESSON 4: HOW CAN WE GET DRINKABLE WATER FROM SEAWATER?

LESSON SUMMARY:

This is the third of three lessons exploring the mixing and separating of liquids and solids. In this lesson children recall their understanding of changes of state from Year 4, Module 1, In a State, and build on the work carried out in Lessons 2 and 3. They use their knowledge of evaporation and condensation to work out how to get materials back from a solution by investigating a real world problem: how to produce drinkable water from seawater, using limited equipment. By the end of this lesson they will have followed or developed their own method and recognised how and why the amount of drinkable water produced varies.

Preparation required: the results of this investigation will only be seen after a few hours have passed, so an early morning start will ensure maximum impact and that children can observe water as it condenses as a result of the evaporation process.

National curriculum links:

Use knowledge of solids, liquids and gases to decide how mixtures might be separated, including through filtering, sieving and evaporating

Learning intention:

To explain the processes of evaporation and condensation and how these might help to produce drinkable water from a plentiful supply of seawater

Scientific enquiry type:

Observation over time

Working scientifically links:

Planning different types of scientific enquiries to answer questions, including recognising and controlling variables

Success criteria:

- I can describe how dissolved material can be separated from a liquid.
- I can follow instructions to set up equipment to produce drinkable water from seawater.
- I can develop my own method to produce drinkable water from seawater.
- I can explain why the amount of water produced varies.

EXPLORE:

Discuss the children's observations of what happened to their solutions from the previous lesson.

Ask: *What do you think happened to the liquid? Is the material left there now the same as you originally put into the liquid?*

Remind them of the work on changes of state and evaporation they did in Years 3 and 4. You might suggest that they have managed to get the solid back from the solution but not the liquid, and that is what they are going to look at today.

Show children the cartoon (Slideshow 1) to set the scene. Explain that an adventurer has been shipwrecked on a desert island. His fresh water supply is running out and he needs to find a way to provide himself with drinkable water while he waits to be rescued. He has very little equipment available, but there is plenty of water – the only problem is, it's sea water!

Ask: *What do you know already that might help you?*

Ask children to talk to a partner to come up with some ideas.

Use the Prompt question cards (Resource sheet 1) to structure the discussion. Collect ideas from children and note them on the interactive whiteboard.

ENQUIRE:

Explain to children that they have limited resources available and only a short time to test their ideas.

Ask: *How could you use what is available in the classroom to set up an investigation?*

In Challenge 1 children follow a suggested method to collect water evaporated from seawater. In Challenges 2 and 3 children develop their own methods and find the best way to speed up the evaporation and condensation process. In Challenge 3 there is an added complexity: how can their classroom-based model be adapted for use by the stranded adventurer to produce as much drinking water as possible?

Challenge 1: Children follow a suggested method to collect water evaporated from seawater and then use word cards to describe the processes taking place

Provide the children with a diagram of how they should set up their test (Resource sheet 2).

Ask: *What is the purpose of the lamp? What difference will it make? Would any water evaporate if the lamp were switched off? How long do you think it will be before water droplets begin to appear on the clingfilm? Where will the water have come from? Could you drink the water or would it taste salty? What will be left when the seawater has gone?*

When the children have set up their test, give them a set of word cards between two (Resource sheet 3) to use to make a sentence to describe the process that is taking place. They should use as many of the words as they need and add others.

Challenge 2: Children develop their own methods to collect water evaporated from seawater

Give the children a selection of materials to consider as they design their test, such as some plastic sheet, some sticky tape, some acrylic sheet, bowls, beakers and other containers to hold the seawater. Prompt the children's thinking during their discussion.

Ask: *How could you speed up the rate of evaporation (in the classroom)? Where will you place the container of seawater so that it will evaporate more rapidly?*

Children might suggest near a radiator or in bright sunlight, depending on the time of the year.

Ask: *Will the water evaporate more rapidly from a shallow wide beaker or a deep narrow beaker? Why?*

When the children have set up their test, ask them to draw a labelled diagram showing the processes that are taking place as the water droplets are produced.

Challenge 3: Children design a method to collect the maximum volume of water evaporated from seawater

Explain to the children that their added challenge is to generate the largest amount of drinkable water in the shortest amount of time. They should plan how the stranded adventurer might do this, bearing in mind that he is on a desert island, and create a model showing how they think he could create his own drinking water generation facility using his limited resources, as described on the adventurer's inventory (Resource sheet 4).

REFLECT AND REVIEW:

Return to the tests after several hours to see whether drinkable water has been produced.

Ask: *Which of your ideas has worked the best? Why do you think that is? What problems were there? In the classroom, where was the best position to set up the equipment? Why? What made a difference to how much water was produced?*

Ask children who completed Challenge 3 to describe their plans for the stranded adventurer. They should explain what equipment he should use and how he should set it up.

Ask: *Where would the stranded adventurer's water generation facility be positioned? Under a tree? Or in full sun?*

Ask Challenge 1 and 2 children to explain why full sun would be best.

Remind children that their stranded adventurer will need to collect enough water to keep him going. For homework, see whether they can find out what the minimum amount of water a day might be for an adult like him, or for a child, in very hot or cold weather.

EVIDENCE OF LEARNING:

Listen to children's responses throughout the lesson. Were children able to follow instructions to set up their equipment to produce drinkable water from seawater? Could they describe the processes involved? Could children develop their own methods and designs to produce drinkable water from seawater? Were they able to explain why the amount of drinkable water produced might vary depending on, for example, position of equipment (full sun or shade, and so on), surface area of container, or whether the wind was blowing? Do children understand that using evaporation and condensation enables the materials that make up solutions to be separated?

MARVELLOUS MIXTURES

Key vocabulary:

material, separate, mixture, sieve, filter, evaporate, solid, liquid, particle, dissolve, soluble, solution, contaminated, impurity, pure, purity

Resources:

Chunky rock salt with impurities, sand, gravel chips and soil; materials to create filter beds, such as felt, wood shavings, sand, insulation fibre, wadding, cotton wool, three 1-litre plastic lemonade bottles pre-cut at neck (these will be used to create filters that can be prepared by children in advance of the lesson – see Resource sheet 1), water, water jugs, selection of sieves and funnels of different sizes

Health and safety:

Wash hands after handling rock salt. Do not taste the salt.

LESSON 5: HOW CAN WE PURIFY MATERIALS?

LESSON SUMMARY:

In this lesson children draw on the work they have done in Lessons 1–4 in order to consolidate their understanding of separating mixtures. They are challenged to develop their own methods to separate pure salt from a rock salt mixture. By the end of this lesson children will have separated the mixture practically and produced a salt solution. They also explore why salt's properties mean that it is useful to industry.

National curriculum links:

Use knowledge of solids, liquids and gases to decide how mixtures might be separated, including through filtering, sieving and evaporating

Learning intention:

To demonstrate and explain how pure salt can be separated from a rock salt mixture, using techniques based on the properties of the materials involved

Working scientifically links:

Planning different types of scientific enquiries to answer questions, including recognising and controlling variables where necessary

Success criteria:

• I can produce pure salt from rock salt using my knowledge of separating mixtures of materials.

• I can describe and explain the process of sieving and filtering mixtures to remove particles of a very small size and suggest a variety of equipment that might be used to do this.

• I can describe and explain how to separate removing solids from solutions using the process of evaporation.

EXPLORE:

Show children a sample of natural rock salt.

Ask: *What do you think this is? How can you find out?*

Explain that it was dug out of the ground. Use the images from Slideshow 1 to discuss the way rock salt is mined and show the children the size of the tunnels.

Ask: *Do you think you could get pure salt from this? How would you do it?*

ENQUIRE:

Remind children of the work they have done on separating materials using sieves, filters and evaporation of liquids to get back the dissolved substances.

Explain that their challenge during this lesson is to use this knowledge to clean up 'raw' rock salt and get some pure salt. Children completing Challenge 1 are supported to develop a method through teacher questioning, while in Challenges 2 and 3 children are expected to work more independently. They can use any of the sieves they made in the last lesson, plus lemonade bottle filter funnels (see Resource sheet 1 for diagram), filter bed materials and plenty of water.

The challenges are presented on the Challenge slides to be displayed on the board, or printed out and placed in the centre of the table.

Challenge 1: Children devise a method, with support, that produces the cleanest salt possible, combining sieving and filtering techniques

Prompt the children, as necessary, to think about the sieving activities that they completed in previous lessons.

Ask: *What equipment will you use to clean up the rock salt? How could you separate some of the mixture before adding water? What sort of sieve will you use to remove those particles? When you add water to your mixture, what particles will you still need to remove by filtering? What material will you add to your filter so that it will remove the tiniest particles?*

Allow the children to complete the process.

Ask: *Can you describe what is left in the filtered solution? How can you separate these materials?*

Ask the children to draw a flow chart with words and images to describe the process of separating the salt.

Challenge 2: Children devise their own method to create the cleanest salt possible, combining sieving and filtering techniques

Ask: *What equipment will you need? How will you set it up? Which material will be the first you separate from the mixture? Why? How will you make sure that the salt solution you are left with is as pure as possible?*

After they have completed the process ask them to describe the mixture in the filtered solution.

Ask: *What is the next step in the process to separate the pure salt?*

Ask the children to write a set of instructions for someone to continue the process.

After they have completed the process, ask them to write a project report to answer the following questions: Can you describe what is left in the filtered solution? Are you confident that any impurities have been removed? How could the purity of the salt be improved?

Challenge 3: Children evaluate their own method to create the cleanest salt possible, combining sieving and filtering techniques

Ask: *What equipment will you need? How will you set it up? How will you judge the success of your system?*

After they have completed the process, ask them to write a project report to answer the following questions: Can you describe what is left in the filtered solution? Are you confident that any impurities have been removed? How could the purity of the salt be improved?

REFLECT AND REVIEW:

Discuss the difficulties of getting pure salt from rock salt. How successful were they? Can they think of one plus for the method they used and one minus, and one thing they could improve?

Show children the images from the salt mines at Winsford again (Slideshow 1). Explain that salt is a very important material because it is used in the manufacturing process of many everyday products. Only 30 per cent of salt is used on or in our food.

For homework, ask children to find out how salt is used in the manufacture of one of the products listed in Resource sheet 2. They should use a variety of appropriate texts, plus the internet, to find out as much information as they can. Can they find out which industry uses the most salt?

EVIDENCE OF LEARNING:

Listen to children's responses throughout this lesson. Can they describe the process of filtering to remove particles of a very small size from a mixture? Are they able to suggest a variety of equipment that they might use to do this? Can they explain that the smaller the solid particles are that they need to separate, the smaller the holes in the filter will need to be? Can they come up with a method to successfully filter a liquid from a solid? Do they recognise that repeating the filtering process, using fresh water and a finer filter, will improve the purity of the salt? Can they explain that the salt can be retrieved from the water solution by allowing it to evaporate? Do they identify any problems with the method they used? Can they give examples of how salt is used in industry?

CROSS-CURRICULAR OPPORTUNITIES:

There are geological elements to this lesson which can be linked to Geography.

Key information:

Salt (chemical name – sodium chloride) gives us chlorine for the chlor-alkali chemical industry – the biggest user of salt. Chlorine-based chemistry provides clean water, soaps and detergents, many medicines, PVC pipes for homes, mobile phones, cosmetics, protective suits for divers and astronauts, digital cameras, flat panel TVs, electron microscopes and solar panels for energy production. Unpurified rock salt is used for salting roads in the winter.

MARVELLOUS MIXTURES

Key vocabulary:

material, compare, contrast, separate, mixture, sieve, filter, evaporate, solid, liquid, gas, powder, particle, dissolve, soluble, solution, suspension, reversible, non-reversible

Resources:

Transparent pint beakers, cheapest vegetable oil, cheapest lemonade (in large quantities), food colour and water droppers, table salt, trays full of sand to stabilise beakers of liquid

Health and safety:

Handle liquids with care. Avoid spilling oil on floors.

Key information:

Oil is less dense than water, so floats on the surface of water (or lemonade, which is mostly water). Dissolving takes place when two materials share a similar chemical property. Salt and water molecules carry positive and negative charges. As salt is mixed into water, the charged water molecules break apart the charged salt molecules. The sodium and chloride molecules mix uniformly with the surrounding water molecules and the salt dissolves in water, forming a mixture. Oil, on the other hand, is made up of chains of molecules that are not charged, so are very different from both salt and water molecules. The salt will not dissolve in the oil, but will in the lemonade.

 ENRICHMENT LESSON 1: WHAT WILL HAPPEN IF WE ADD A SPRINKLE OF SALT TO A COMBINATION OF LIQUIDS?

LESSON SUMMARY:

In this lesson children explore what happens when oil and lemonade mix and the effect of a sprinkle of salt on the combination. By the end of this lesson they will have made predictions and careful observations, and some will be able to explain their understanding of what happens using their science subject knowledge and suggest how variables might be changed in further investigations.

Preparation required: The process of creating the Oily Hubbub would be ideal to video (or photograph). Watching each step for a second time will allow some to think again and improve the quality of their responses.

National curriculum links:

Understand that some materials will dissolve in liquid to form a solution, and describe how to recover a substance from a solution

Learning intention:

To describe and explain observations of what happens to a sprinkle of salt as it is added to a mixture of lemonade and oil

Scientific enquiry type:

Observation over time

Working scientifically links:

Reporting and presenting findings from enquiries, including conclusions, causal relationships and explanations of and degree of trust in results, in oral and written forms such as displays and other presentations

Success criteria:

- I can describe what happens when a sprinkle of salt is added to a mixture of oil and lemonade.
- I can suggest explanations for what happens after each step of the enquiry.
- I can suggest further investigations that I could carry out by changing the variables in my investigation.

EXPLORE:

Refer back to previous lessons. Ask children to work in twos or threes to create a concept map using a selection of the Key word cards as a starter (Resource sheet 1).

Ask: *What connections were you able to make? Are there any technical words that you can't remember? We know that certain solids dissolve in water – which were they? But do those solids dissolve in all liquids?*

From their work in previous lessons children will explain that solids which dissolve in water do not dissolve in all liquids – for example, salt does not dissolve at all in oil.

ENQUIRE:

Explain to children that during this lesson they are going to work in groups to observe what effect adding a sprinkle of salt has on a combination of the liquids oil and lemonade. They will need to observe closely at every stage and use their subject knowledge to come up with explanations for what they see taking place.

Provide each group with a transparent pint beaker, a third-filled with oil.

Ask children to predict what they think will happen if they add a similar volume of lemonade to the oil:

Ask: *Will the oil and lemonade mix together completely or not? Will the oil stay at the bottom of the beaker? What will happen to the lemonade? What will happen to the bubbles of gas (carbon dioxide) in the lemonade?*

Add the same volume of lemonade and observe what happens.

Ask: *Was your prediction correct? What happened? What do you notice about the oil and lemonade? Why does the oil float on top of the lemonade? What is your explanation?*

Use a water dropper and place four or five drops of food colour on the surface of the oil.

Ask: *What happens to the droplets of oil? What shape are they? Do they change as they move through the oil? Do any of them mix with the lemonade? What happens to the food colour? What is your explanation?*

The next step is to add a sprinkle of salt: a small quantity only, beginning with a pinch. Ask children to describe what they think might happen to the salt.

Ask: *Do you think it will dissolve in the oil or the lemonade? If it dissolves in the lemonade, what effect will that have on the bubbles?*

Sprinkle a pinch of salt onto the surface of the oil and then wait.

Ask: *What do you notice? Why is a rush of bubbles produced as the salt reaches the lemonade? What might be the cause? What is your explanation?*

The challenges are presented on the Challenge slides to be displayed on the board, or printed out and placed in the centre of the table.

Challenge 1: Children use a grid to describe their observations

Ask the children to record their observations on the Describing observations grid (Resource sheet 2). Prompt them to think carefully about the scientific vocabulary they are using and to be as precise as they can in describing their observations and explanations.

Challenge 2: Children communicate their explanations of what happened using technical vocabulary

Ask the children to think through the steps of the enquiry and report their findings. They can do this however they wish but must include: what they observed; what was happening to the materials; and their explanation for what they saw using technical vocabulary. Remind them that there are three main steps they need to comment on: adding lemonade to oil; adding food colour; adding a sprinkle of salt. They can refer to the concept map to remind them of the scientific vocabulary.

Challenge 3: Children communicate their explanations of what happened

Explain to the children that they need to find an effective way to communicate their explanations of what happened. They can do this however they wish, but must comment on what they observed and the causes of what happened, and provide an explanation using technical vocabulary.

REFLECT AND REVIEW:

Bring children back together to share their observations and explanations of what happened at each step of the process. If possible, look with children at your pre-recorded video or any photographs you took to prompt their thinking. Summarise their responses on the interactive whiteboard step by step, ensuring that explanations use key technical vocabulary appropriately.

Show children Slideshow 1, which describes what some other children said when they did the investigation in their class. There are lots of mistakes: children should be able to point these out and explain what actually happens, using the correct vocabulary.

Ask: *If you could change the investigation in any way you like, to see what difference it would make, what would you do?*

Prompt children, if necessary, to think about different solids they could add – would the same thing happen with sugar or currants? Encourage them to think about the liquids they could use, or the volumes of liquid, or quantities of solid. Encourage them to shape their ideas into questions.

EVIDENCE OF LEARNING:

Listen to children's responses throughout the lesson. Do they describe their observations in sufficient detail and using correct technical vocabulary? Are they able to offer an explanation of what happens after each step of the enquiry, using their science knowledge to help them make sense of what they have observed? Are they able to suggest further investigations they could carry out, by changing the variables investigated (for example, changing quantities, using cola instead of lemonade, or fizzy water, or a still liquid, adding sugar or another solid instead of salt)?

MARVELLOUS MIXTURES

Key vocabulary:

material, compare, contrast, separate, mixture, sieve, filter, powder, particle, dissolve, soluble, solution, suspension, contamination, microbes, bacteria, purify, drinkable, sterilise

Resources:

Containers of 'contaminated' water – a soupy mixture containing as nasty a mix as you like: oil, soil, sand, pebbles, leaves and twigs, bits of plastic, and so on, and some pond water for children completing Challenge 3. Sieves with different grade mesh, funnels of different size, filter papers, buckets, bowls, plastic sheeting, mop and bucket. Material collection for making filter beds – fine sand, gravel, wadding, felts and other thick fibrous fabrics, foil food trays, microscopes

Health and safety:

Ensure children handle the 'contaminated' water with care and wash hands thoroughly afterwards. They must NOT try to drink any of the water. Cuts on hands should be covered with waterproof plasters.

Key information:

Reassure children that cholera is no longer a problem in this country. It is a disease caused by contaminated water and is usually found in parts of the world where a major natural (or other) disaster has happened, in the aftermath of massive floods or tsunamis, or where sanitary conditions are poor

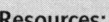

ENRICHMENT LESSON 2: HOW CAN WE CLEAN UP CONTAMINATED WATER?

LESSON SUMMARY:

In this lesson children use their knowledge of separating mixtures to help them solve a real world problem. By the end of this lesson they will have identified, by testing, the most effective methods to use to clean up contaminated water. Children will also be able to explain that water which appears clean to the naked eye may still be undrinkable.

Preparation required: Make a supply of contaminated water. If possible, add some pond water, from an unpolluted source, to the mixture examined by Challenge 3 children, so they have something to see when using their microscopes. This lesson will take longer than an hour and you may wish to extend it into a larger-scale project where children investigate how cholera was eradicated from British cities.

National curriculum links:

Use knowledge of solids, liquids and gases to decide how mixtures might be separated, including through filtering, sieving and evaporating

Learning intention:

To investigate a variety of methods that could be used to clean up contaminated water

Working scientifically links:

Reporting and presenting findings from enquiries, including conclusions, causal relationships and explanations of and degree of trust in results, in oral and written forms such as displays and other presentations

Success criteria:

- I can identify different processes that could be used to clean up contaminated water.
- I can describe what methods, materials and equipment to use to clean the water effectively.
- I can explain that clear water is not necessarily clean and suggest different ways that the water could be purified, so that it is safe to drink.

EXPLORE:

Show children Slideshow 1, about a disastrous natural event that happened several years ago in Haiti, a very poor country in the Caribbean. An initial earthquake was made very much worse by heavy rains and flooding. Water and food supplies became contaminated and there was a serious outbreak of the disease cholera. Ask children questions about the information displayed.

Ask: *What was the sequence of events? Where did the people live after the earthquake? What do you think conditions would have been like? What problems might they have faced? Has anybody heard of cholera before? How does the disease spread? What are the symptoms?*

ENQUIRE:

Explain to children that their challenge during this lesson is to use what they have learned already about separating mixtures to help them come up with some methods of cleaning water that could be recommended to the people of Haiti. The country is poor and resources are few, so the simpler (and cheaper) the methods the better. Children should choose a method to investigate how to clean water: either a practical investigation to construct, and test, a filter system from sieves and filters, or a filter bed system from different sized stones and sand; or a research activity using secondary sources of information.

The challenges are presented on the Challenge slides to be displayed on the board, or printed out and placed in the centre of the table.

Challenge 1: Children use their knowledge of sieving and filtering to come up with a sequence of separating processes they could use to clean dirty water

Provide a bowl or bucket of contaminated water for each group and a selection of sieves, filters and filter papers. Explain to children that they are going to decide on a method for cleaning up their contaminated water sample. They are to photograph each step in their sequence and produce a presentation later, giving detailed instructions about how to clean up the water supply.

Ask: *What will you do to clean the water? What will you do first? What next? What equipment will you use? Will a sieve with big holes take out all the tiny particles? How will you do that?*

Ask the children to discuss the sequence of their process and decide what to use.

Give each group a process planner (Resource sheet 1) to write notes about what they will do and to identify the photos they need to take.

Challenge 2: Children create the most effective filter using a variety of materials to create filter beds to clean up a contaminated water supply

Explain to the children that they need to use what they know about filters and how they work to create a series of filter beds that could be used to clean up the contaminated water. Provide them with a selection of materials to choose from and a number of suitable containers to use as the structure for their filter beds, such as large foil food trays with holes punched in them. Filter bed instructions on Resource sheet 2 model the process of making one.

Ask: *How will you remove the largest particles from the water? What types of materials might you put into your filter beds? How will you remove the tiniest particles? How many times will you need to filter the water?*

They should plan how they will filter the water and work together to achieve as clean water as possible. Ask them to make notes about the sequence of filtering and materials used that are most effective, as these will form the basis of their recommendations.

Challenge 3: Children research methods that might be used to purify water and suggest the best of these as a final stage to the water cleaning process.

Ask: *How clean will the water be after sieving and filtering? Do you think the water will be safe enough to drink? What might cause further problems to healthy as well as sick people? How do we filter water so that it is clean to drink? What are the water filters used in water systems, water jugs, in houses and swimming pools made from? What other treatments could be used?*

The children should use microscopes to check for tiny creatures not visible to the naked eye. They should research ways in which water can be purified and make suggestions. For example, they might know that water in swimming pools is treated with chemicals to make it safe or that boiling water kills most bacteria.

They should summarise their findings in a 'safe to drink' fact sheet.

> **Key information:**
> Microscopes are available on loan (for a term, available in kits, including eight microscopes and a camera) from the Royal Microscopical Society (rms.org.uk/outreach/activitykit/How+do+I+get+one)

REFLECT AND REVIEW:

Children should summarise the findings from the investigations they have carried out. Ask children who have completed Challenge 1 or 2 to describe what seemed to work best.

Ask: *What were the most effective methods? Was the water completely clean? Was it clear or was it still slightly cloudy? Could there have been solids dissolved in it? How could you test it to see if it has solids dissolved in it? Do you think the water would be safe enough to drink?*

Children who completed Challenge 3 can then describe what they learned.

Ask: *How confident would you be that the water is safe to drink? What should you do next to make the water safe? What methods could you use?*

To complete this enrichment lesson, and extend children's thinking still further, give each group a set of large Ranking cards (Resource sheet 3). Explain that these solutions have been proposed to help improve the situation in Haiti. Which would they recommend and why?

Children should read, discuss, then rank the cards from best to worst solution. They must make decisions, but also be prepared to justify what they are suggesting, so they need to think carefully about all they know that could help them decide.

EVIDENCE OF LEARNING:

Listen to children's responses throughout the lesson. Can they identify different processes that are useful in cleaning up contaminated water, based on their knowledge of separating mixtures from previous lessons? Can they transfer their knowledge of methods of separating materials to this context and identify what materials or equipment to use and how to achieve clean water? Are they able to explain that clear water is not necessarily clean and could some of them suggest different ways that the water could be purified, so that it is safe to drink?

Thanks to Sarah Bell for allowing us to reproduce this activity.

MATERIALS: ALL CHANGE!

INTRODUCTION

In this module children develop their knowledge and understanding of changes to materials. They will recognise that some changes such as melting, evaporation and dissolving are reversible while other changes are non-reversible, including burning and production of rust or a gas as the result of a chemical reaction. Children use specific scientific vocabulary as they describe, explain and communicate their understanding of how materials change. Working scientifically, children observe and compare changes that take place over time in a variety of different contexts, such as when making toffee, rusting metals or burning a candle. They plan and carry out fair tests to investigate more systematically non-reversible changes that they observe. They use a variety of ways to report and present their findings to an audience.

This module, together with Modules 3, 4 and 5, builds on earlier learning that began in Key Stage 1 and then continued in Year 4. There, children compared and grouped materials according to whether they were solids, liquids or gases and learned about changes of state that take place when materials are heated or cooled. This series of modules offers the final Chemistry-related learning for children in Key Stage 2. It provides teachers with ample opportunities to assess children's progress against the programmes of study.

Key vocabulary:

material, change, compare, contrast, solid, liquid, gas, change of state, dissolve, melt, reversible, non-reversible, mixture, powder, particle, tablet, bubbles, carbon dioxide, change, reaction, inflate, rust, oxidise, oxygen, corrode, tarnish; types of metal: iron, steel, chromium, tin, zinc; boil, vapour, fuel, heat, burn, burning, flammable, flame, melts, solidifies, candle, wick, wax

FACT FILE:

Reversible changes

Reversible changes are those in which the fundamental composition of the materials involved remains unchanged, and that by altering the conditions it is possible to return the materials to their original state. Changes of state (such as freezing, melting, evaporating, boiling and condensing) are reversible changes brought about by changes in temperature. In some cases the material may look slightly different when it is returned to its original state, but it remains the same material.

Dissolving is another example of a reversible change. The material that is dissolved, such as salt (called the solute), in another material, such as water (called the solvent), can be recovered by separating the two materials. For example, the water can be allowed to evaporate leaving the salt in a dish and the water can be collected by letting it condense on a cold surface.

Non-reversible changes

Non-reversible changes occur when materials react to produce new products which cannot be easily turned back into the original materials. For example, sodium bicarbonate, more commonly known as **bicarbonate of soda**, will produce **carbon dioxide** gas if it is heated or mixed with weak acids. Other solids behave similarly, including effervescent indigestion tablets, vitamin C tablets and baking powder. However, these changes must be distinguished from the bubbles in a fizzy drink which are caused by carbon dioxide that has been dissolved in the liquid under pressure. When the lid is removed from the fizzy drink, the pressure is released and the bubbles rush to escape.

Bubbles in different materials

Self-raising flour is made from plain flour and baking powder. The proportions of flour to baking powder are adjusted so that 'light' cakes can be baked. Bubbles of carbon dioxide are trapped in the cake mix. As the mixture is heated these bubbles expand, causing the cake to rise. The mixture sets around the gas bubbles. If the bubbles are too large there will be large holes in the cake.

Bread-making uses yeast, which is a living organism that produces carbon dioxide by a series of non-reversible reactions as it grows. The bread dough rises and, on baking, the gases expand still further, adding lightness to the baked bread.

Pumice stone is formed when volcanoes erupt. Lava with gases in it froths and then cools to form the rock, which is very light and can float on water (if dry).

Expanded polystyrene is a type of plastic. It is formed at high temperatures when hot air or water vapour is forced into the plastic. Bubbles form as a result, which increases the insulating properties of the material significantly.

Rusting is another example of a non-reversible change. Almost all metals **rust** or **oxidise** to an extent.

When metals containing iron are exposed to the air, the iron combines with the oxygen and water to form a reddish coloured solid called rust. The rust flakes away from the iron, leaving more iron exposed to the air and so the process continues.

There are several ways to prevent iron and steel rusting. Some of these work because they stop oxygen or water reaching the surface of the metal:

• oiling – for example, bicycle chains

• greasing – for example, nut and bolts

• painting – for example, car body panels

Iron and steel objects may also be covered with a layer of another metal. Food cans, for example, are plated with a thin layer of tin. Galvanising is when an iron or steel object is coated in a thin layer of zinc.

Burning

When materials **burn**, a chemical change takes place. For burning to happen, three things must be present: **fuel**, **oxygen** and **heat**. Heat and light are given out during the change.

When a lit match is placed near to the wick of a candle, the wax on it melts and evaporates into a gas which then ignites producing new gases, heat and light. The heat from the wick burning slowly melts the wax at the top of the candle. The molten wax is soaked up by the wick and evaporates into a gas, which burns to keep the candle alight.

Common misconceptions:

• Children may think that certain solids dissolve in liquids, not recognising that a chemical change has taken place, producing carbon dioxide gas as a result, such as sodium bicarbonate or a vitamin C tablet in water.

• They may think that carbon dioxide bubbles released from a bottle of lemonade when the lid is unscrewed indicate a chemical reaction. They may struggle to understand that the gas released has been dissolved in the liquid and is released because of the reduction of pressure.

• They may think that only the wick of a candle burns, not recognising that the wax melts and vaporises and the gas burns, ignited by the candle flame.

MATERIALS: ALL CHANGE!

LESSON 1: ARE THE CHANGES THAT HAPPEN AROUND US REVERSIBLE OR NON-REVERSIBLE?

Key vocabulary:

material, change, compare, contrast, solid, liquid, gas, change of state, reaction, dissolve, melt, reversible, non-reversible

Resources:

Small bottles of lemonade, shaving foam canisters, salt, water, chocolate buttons, beakers and small plates, paper towels

Key information:

Some materials, when heated or cooled, or when they interact with other materials, change without resulting in new products being formed. By reversing the conditions or separating the mixture of materials, the original substances can be recovered. These are known as reversible changes and include changes of state and dissolving.

Other materials, when heated or brought together, will react and produce new materials as products of the reaction. These are known as non-reversible (irreversible) changes and include burning and other chemical reactions.

LESSON SUMMARY:

This lesson builds on learning about states of matter from Year 4, Module 1, In a State and also from Year 5, Module 5, Marvellous Mixtures, where children investigated how some materials could be mixed together and then separated. In this lesson children begin to explore how materials change when they are brought together in different ways. They identify types of changes and group them according to whether they think the change could be reversed, and then according to the conditions needed to bring about the change.

By the end of this lesson children will be able to use key vocabulary effectively to communicate their scientific understanding of reversible and non-reversible changes.

National curriculum links:

Demonstrate that dissolving, mixing and changes of state are reversible changes. Explain that some changes result in the formation of new materials and that this kind of change is not usually reversible, including changes associated with burning and the action of acid on bicarbonate of soda

Learning intention:

To describe different changes in materials when they are brought together and to be able to recognise them as reversible or non-reversible

Scientific enquiry type:

Grouping and classifying

Working scientifically links:

Reporting and presenting findings from enquiries, including conclusions, causal relationships and explanations of and degree of trust in results, in oral and written forms such as displays and other presentations

Success criteria:

- I can identify familiar examples of materials changing state.
- I can describe what causes some materials to change.
- I can explain, using examples, that some changes are reversible while others are non-reversible.

EXPLORE:

Show children the Busy kitchen image (Interactive 1), which shows a kitchen scene featuring materials in different stages of change – a boiling kettle, toast in the toaster, some cans of drink with beads of condensation on the outside, tall glass beakers with ice cubes, an open bottle of fizzy pop, a freezer with its door open, a frying pan on the stove with egg cooking, bread cooking in the oven, melted butter on toast, milk in a jug to pour on cereal, a teapot with a sugar bowl and mugs, and a box of eggs dropped and broken on the floor.

Ask children to talk about the cartoon with their partner. Encourage them to identify examples of materials and whether they have (or are about to) change.

Ask: *What is changing? How is it changing? Do you think you can get back to the original materials?*

Encourage children to give reasons for their answers. Click on the interactive to reveal whether the changes identified are reversible or non-reversible.

ENQUIRE:

Introduce the terms 'reversible' and 'non-reversible' changes and explain to children that they will be finding out more about different kinds of material changes during this lesson and the rest of the module.

Challenge 1: Children identify whether familiar changes are reversible or non-reversible and explain their answers

Provide the children with Name that change! (Resource sheet 1), one between two, and challenge them to 'name that change' as reversible or non-reversible. Ask them to decide yes, no or maybe so (if they are not sure) in a given number of contexts. They can add some more examples, perhaps taken from the Kitchen chaos cartoon discussed in the Explore section.

Key information:

At this stage it is unlikely that children will be able to classify all the changes they recognise, but this should improve as the module progresses.

Key information:

Some changes are difficult to classify on simple observation; for example, some children will identify opening the lemonade bottle as non-reversible, because the carbon dioxide gas escapes from the drink. In fact the carbon dioxide is dissolved in the lemonade under pressure, so it is actually reversible, but with difficulty.

Ask: *Do you think it is possible to get the original materials back? Can you explain why you think a particular change is reversible or non-reversible?*

Challenge 2: Children use a graphic organiser to compare and contrast reversible and non-reversible changes

Provide pairs of children with an enlarged copy of the Compare and contrast grid (Resource sheet 2). Tell them to discuss the differences between the materials listed and fill in the boxes with their ideas, as best they can. Prompt their thinking, once they have brainstormed their ideas, by providing examples of the materials for the children to explore. Encourage them to expand on their initial responses, recording any additional observations and descriptions on their grids.

Ask: *Which types of change did you find examples of? Is it possible to get the original materials back? Do you think the change is reversible or non-reversible? Were there any you were not sure about?*

Challenge 3: Children create a graphic organiser to compare and contrast reversible and non-reversible changes

Provide pairs of children with an enlarged blank copy of the Compare and contrast grid (Resource sheet 2). Ask them to choose some changes they noticed in the Explore activity, plus some they know from their own experiences, and complete the grid. They need to include a range of different changes, so not all melting from solid to liquid.

Ask: *Which types of change did you identify? Is it possible to get the original materials back? Do you think the change is reversible or non-reversible? Were there any you were not sure about?*

REFLECT AND REVIEW:

Give the children, again working in pairs, the Odd one out cards to discuss (Resource sheet 3). Ask them to decide, between themselves, which is the odd one out and why. Remind them that there are no 'right' answers, because it depends on the criteria they choose; for example, cheese, sugar and salt. Cheese might be odd because it does not dissolve, but the others do; sugar could be odd because you might put cheese and salt in an omelette, but not sugar. When they are happy with an odd one out they have chosen, they should share it with another pair, swap cards, and begin the process again. Allow several swaps to take place.

Ask: *Did anybody come up with an odd one out that involved a reversible or a non-reversible change?*

Discuss jelly, fudge, ice, sherbet and flour in terms of reversible and non-reversible change.

Ensure throughout that the children are using scientific vocabulary, such as melting, dissolving and evaporating to describe processes correctly, and showing understanding of reversible and non-reversible changes appropriate at this stage in the module.

Key information:

At this stage children will have greater experience of reversible change and may only have limited awareness of non-reversible change. This is built on during the rest of the module.

EVIDENCE OF LEARNING:

Listen to children's responses throughout this lesson and consider the recorded outcomes of the three challenges. Are they able to identify familiar examples where materials are changing state, such as a kettle boiling, baking, frying or making ice in a freezer? Can they describe how specific materials have changed, using scientific vocabulary such as melting, evaporating, dissolving or freezing? Can they identify and explain some changes as reversible while others are non-reversible?

MATERIALS: ALL CHANGE!

LESSON 2: HOW MUCH GAS CAN BE PRODUCED BY NON-REVERSIBLE CHANGE?

LESSON SUMMARY:

In this lesson, as an example of a non-reversible change, children explore a variety of solids and liquids that react chemically when they are mixed. By the end of this lesson they will have identified the best combination of materials to use to inflate different containers such as plastic disposable gloves, balloons and plastic bags of varying capacities.

Key vocabulary:

material, mixture, solid, liquid, gas, powder, particle, tablet, bubbles, inflate, carbon dioxide, reversible, non-reversible, change, reaction

Resources:

Disposable latex gloves; solids: bicarbonate of soda, tartaric acid, baking powder, effervescent vitamin C tablets, effervescent indigestion tablets; liquids: water, white vinegar, lemon juice; small beakers, disposable cups, plastic teaspoons, milk bottles or cartons, small pop bottles

Key information:

Sodium bicarbonate, more commonly known as bicarbonate of soda, will produce carbon dioxide gas if it is mixed with vinegar or some other liquids. This is an example of a chemical reaction that is a non-reversible change. Other solids behave similarly, including effervescent indigestion tablets, vitamin C tablets and baking powder.

Health and safety:

Warn children not to put substances into their mouths (see Be Safe! section 8).

National curriculum links:

Explain that some changes result in the formation of new materials, and that this kind of change is not usually reversible, including changes associated with burning and the action of acid on bicarbonate of soda

Learning intention:

To compare the quantity of carbon dioxide produced by combinations of solids and liquids that react chemically

Scientific enquiry type:

Carrying out comparative and fair tests

Working scientifically links:

Using test results to make predictions to set up further comparative and fair tests

Success criteria:

- I can test different combinations of materials to find out which produces the most gas.
- I can vary a mixture so that sufficient gas is produced to inflate a plastic glove.
- I can use the terms 'physical' and 'chemical' change correctly.

EXPLORE:

Remind children that in the last lesson they identified different types of changes: reversible changes (dissolving, evaporating and freezing) where the original material can be retrieved; and non-reversible changes that produce a new material.

Explain that in this lesson they are going to investigate a non-reversible change.

Show them some bicarbonate of soda and white vinegar.

Ask: *What do you think they are? Why?*

When several ideas and reasons have been given, allow one child to taste each one and then ask again: What do you think they are? Establish that the bicarbonate of soda is not salt or sugar and that the vinegar is not water.

Add the vinegar to the bicarbonate of soda and ask children to observe very carefully what happens. Establish that bubbles are given off containing a gas. Explain that a chemical reaction has taken place.

Show them a demonstration of how to inflate a disposable plastic glove using the gas produced by a mixture of bicarbonate of soda and white vinegar. Use a small fizzy drink or water bottle, plastic milk carton or similar and elastic bands to fasten the glove to the bottle. Place about 50 ml of white vinegar in the bottle and about two tablespoons of bicarbonate of soda in the glove. Take care to shake the powder into the fingers of the glove. Attach the glove to the rim of the bottle with the elastic bands, without tipping any powder into the bottle at this stage. When everyone is ready, shake the powder from the glove fingers into the bottle and observe carefully what happens. Time how long it takes for the glove to stand upright, waving at the class!

Ask: *What made the glove expand? Where did the gas come from? How long did it take for the glove to fully inflate? What difference would it make if we used less or more bicarbonate, or less or more vinegar? What about other combinations of materials? Would there be the same reaction with water?*

ENQUIRE:

Provide children with a selection of solids and liquids and some small beakers or disposable cups to use for their initial tests. Explain that they are going to test different combinations of materials to find out which is the best mix to inflate their plastic glove.

In Challenge 1 children are supported by a 'potion checklist' specifying the materials and proportions they should test. In Challenge 2 children explore a wider range of materials with the aim of coming up with a 'perfect' combination: one that will best inflate the plastic glove. In Challenge 3 children investigate how to create a reaction to inflate different balloons and plastic bags, varying the quantities that they use to achieve the best mix.

Challenge 1: Children test a selection of solids and liquids to find out which is likely to provide the best mix to inflate their plastic glove

Provide the children with a limited selection of materials and a Potion checklist (Resource sheet 1) to help them structure their test.

Allow the children to complete their tests.

Ask: *Which of the potions you tested produced the most gas? Do you think the mixture produced enough to inflate your glove?*

The children should then test out their mixture using a glove, increasing the amounts of solid and liquid as necessary.

Challenge 2: Children test a selection of solids and liquids to find out how to make the best mix to inflate their plastic glove, adjusting proportions of solid and liquid to ensure the reaction produces sufficient gas

Provide the children with a selection of materials and liquids to use to make 'potions'. If necessary, prompt them with some guidance as to the proportions between solid and liquid (1:2 solid to liquid), but encourage them as much as possible to experiment using different amounts of the materials available. Give them a Potion recipe card (Resource sheet 2) to stimulate their ideas.

Allow the children to complete the tests.

Ask: *Which of the potions you tested produced the most gas? Do you think the reaction produced enough to inflate your glove? Or would you need to use more solid, more liquid or both?*

The children should then test out their mixture in a glove, adjusting the proportions of solid and liquid as necessary.

Challenge 3: Children evaluate basic potion recipes and then vary the quantities to inflate different sized bags and balloons

Inform the children that your potion recipe says that 30 g of bicarbonate of soda and 50 ml of white vinegar should be enough to inflate a disposable plastic glove. If this is the case (and they will need to check), how much of each material will they need to inflate other things? Provide the children with a variety of balloons of different sizes and shapes, plus some plastic bags of different sizes, to test.

Ask: *Was the potion recipe effective? Was the volume of gas produced sufficient to inflate the glove? What combination of bicarbonate of soda and vinegar produced the most gas? What was the largest thing you managed to inflate? Are any other materials more effective at producing this reaction? For example, would effervescent indigestion tablets, with water or vinegar, produce a greater quantity of gas?*

REFLECT AND REVIEW:

Ask children to share which potion they found to be most effective at inflating the plastic glove. Explain that you need to inflate some balloons. What would be the best way for you to do it? Display Interactive 1.

Explain that there are a number of options on screen, such as using a foot pump or a mix of chemicals or heating the air up, and that together you are going to sort them into reversible or non-reversible changes.

Ask: *Are there examples you can think of where things are inflated using these methods? What changes are happening in each case? Are they reversible or non-reversible?*

EVIDENCE OF LEARNING:

Listen to children's responses throughout the lesson and look at their records of their tests. Are they able to test different combinations of materials independently to find out which produces the most gas? Are they able to vary their mixtures of materials so that sufficient gas was produced to inflate a plastic glove? Are they able to explain the differences between a chemical, non-reversible and a physical, reversible change?

MATERIALS: ALL CHANGE!

 ## LESSON 3: HOW LONG DOES IT TAKE FOR IRON NAILS TO RUST?

LESSON SUMMARY:

This is the first of two lessons exploring another example of a non-reversible change: rusting. By the end of this lesson children will have set up an investigation to observe the changes that take place when some metals are exposed to the air or water.

The changes that children investigate during this lesson will not be apparent for about two weeks. The enquiry should be set up during the lesson, changes monitored over time and the results considered after an appropriate period has passed. These results will be collated and discussed during Lesson 5 (which should take place around two weeks after this lesson).

National curriculum links:

Explain that some changes result in the formation of new materials and that this kind of change is not usually reversible, including changes associated with burning and the action of acid on bicarbonate of soda

Learning intention:

To plan and set up an observation over time to investigate the conditions required for iron to rust

Scientific enquiry type:

Observing over time

Working scientifically links:

Planning different types of scientific enquiry to answer questions, including recognising and controlling variables where necessary

Success criteria:

- I can recognise that metals corrode; for iron and steel this is called rusting.
- I can describe the process of rusting in iron and steel.
- I can set up an observation over time.
- I can explain why the amount of rusting and the time taken to rust can vary according to the conditions.

EXPLORE:

Show children Slideshow 1, Lucy's birthday bike. Establish whether children have noticed rust on cars or other items. Do they know what it looks like? What do they think it is? Do they know what causes it? Why do people try to prevent it?

Ask: *What do you think of the advice the children are giving Lucy? Do you agree or disagree with what they are saying? What do you know already that can help you decide whether the advice is useful or not? Is what Granddad says true? Are there other ways to prevent metal rusting?*

Explain that some metal parts are painted and some need oiling or grease to stop them rusting.

Ask: *What about a non-metal bike? What other materials can bikes be made from?*

Remind them of the lesson in Module 4, Enrichment Lesson 1, Are all bikes the same?, where they learned about bikes made from different materials.

ENQUIRE:

Explain to children that during this lesson they will be setting up an observation over time to find out what materials rust and what affects rusting.

Challenge 1: Children set up an observation over time to explore how quickly an iron nail will rust, then make and justify an initial prediction and record their findings

Ask the children, working in small groups, to set up four beakers (or transparent jars with lids) for their test. Provide them with a choice of different liquids (water, salt water, cooking oil, lemon juice, white vinegar and lemonade) in which to place their nails, but instruct them to include one nail in a beaker without a liquid; this will act as a simple control for comparison. Ask the children to think carefully before choosing the liquids to test.

Ask: *Will all the nails turn rusty? Which do you predict will rust first? How will you measure the amount of rust?*

Key vocabulary:

material, compare, contrast, solid, liquid, gas, rust, oxidise, oxygen, corrode, tarnish; types of metal: iron, steel, chromium, tin, zinc; reversible, non-reversible, change of state

Resources:

Iron nails, metal paint, paint brushes, petroleum jelly (or similar thick grease), oil, salt, lemon juice, vinegar, lemonade, water, plastic disposable beakers (transparent), clingfilm, objects made of metal, including washers, key, spoons, copper wire, aluminium foil, tin can, zinc and copper nails

Health and safety:

Metal paint is not removed by washing with soap and water. Take great care to avoid spills and keep quantities to a minimum, as only a small amount is required to paint a nail or two.

Key information:

Almost all metals corrode to some extent when they come into contact with air over a long period of time. This results from a chemical process called oxidation. If the metal contains iron, the iron reacts with oxygen in the presence of water to form a reddish coloured solid called rust. The rate of rusting can be slowed down by treating the iron in different ways, including painting.

Remind them to cover the beakers.

Provide a Recording grid (Resource sheet 1) on which they can record their initial prediction and their observations over time. Prompt the children to think about their predictions again by giving them some Think cards (Resource sheet 2). If the children want to alter their predictions as a result, ask them to use a different colour pen and write the change next to their initial thoughts.

Challenge 2: Children set up an observation over time to find out whether the rusting process can be slowed down or prevented

Provide the children (working in twos or threes) with a choice of materials to use to protect four iron nails, such as oil, petroleum jelly, butter and paint, and allow them to test their own ideas as well. Explain that they need to plan what observations or measurements to make, and how often and how to record their initial predictions and the observations they make. Prompt them to include a 'control' nail without a protective coating in their test and remind them, if necessary, that they are changing the coating of the nail so will need to keep everything else the same, including the liquid the nails are placed into.

Challenge 3: Children observe over time whether metals other than iron rust and whether other materials rust

Provide the children with a selection of objects made of metal to test, including an iron nail and copper and zinc nails, and other materials such as plastics and rubber. Ask them to set up and record an observation over time, recording also their initial predictions.

Ask: *Do you think all metals rust? Are there differences between the ways they corrode?*

REFLECT AND REVIEW:

Watch the video extract with children (Video 1).

Ask: *Why was painting the bridge a 'never-ending' task? What is the problem with using iron/ steel to make a bridge like the Forth Railway Bridge? What is it made of? What properties make iron useful for making a bridge? What would cause it to rust more quickly than it would in the middle of a city? How was it protected in the past? What is better about the new treatment? Why is painting the Forth Railway Bridge no longer a never-ending task?*

Iron and steel objects may also be covered with a layer of another metal. Food cans, for example, are plated with a thin layer of tin. Galvanising is when an iron or steel object is coated in a thin layer of zinc. Video 1 shows how the technology for preventing rusting has changed.

EVIDENCE OF LEARNING:

Listen to children's responses throughout the lesson, taking particular note of any predictions they make and their reasons, and as they plan for making their observations over time. Can they make a prediction about whether or how much a material would rust and explain their reasons? Are they able to describe the process of rusting? Do they identify rusting as a non-reversible change that requires water and air to take place? Can they set up an observation over time, deciding what observations to make and at what intervals?

MATERIALS: ALL CHANGE!

 ## LESSON 4: WHAT HAPPENS WHEN A CANDLE BURNS?

LESSON SUMMARY:

In this lesson children observe and discuss the changes involved in burning a candle, recognising that there are reversible and non-reversible changes involved in the process. By the end of this lesson they will understand what a flame needs in order to burn, and how the time a candle stays alight depends on the amount of air (oxygen) available and the supply of material that is being burnt.

Key vocabulary:

material, compare, contrast, solid, liquid, gas, vapour, reversible, non-reversible, fuel, oxygen, heat, burn, burning, flammable, flame, melts, solidifies, candle, wick, wax

Resources:

Candles (see below), metal containers filled with sand, glass jars of varying size, paper, pencils, digital camera, mini whiteboards

Health and safety:

Candles should be placed in sand inside a metal tray. Use chunky candles or nightlights, which will not tip over easily (see Be Safe! section 16).

Key information:

A series of things are happening when a candle burns. When a lit match is placed near to the wick of a candle, the wax on it melts and the wax turns into a gas and then ignites, producing heat. The heat from the burning slowly melts the wax at the top of the candle. The molten wax is soaked up by the wick, turns into a gas and ignites, thus keeping the candle burning.

Key information:

When materials burn, a chemical change takes place. For burning to happen, three things must be present: fuel, oxygen and heat. Heat and light are also given out during the change.

National curriculum links:

Explain that some changes result in the formation of new materials and that this kind of change is not usually reversible, including changes associated with burning and the action of acid on bicarbonate of soda

Learning intention:

To investigate the changes involved in a candle burning

Scientific enquiry type:

Observation over time

Working scientifically links:

Reporting and presenting findings from enquiries, including conclusions, causal relationships and explanations of and degree of trust in results, in oral and written forms such as displays and other presentations

Success criteria:

- I can observe changes involved in burning a candle.
- I can describe the reversible and non-reversible changes that take place.
- I can identify the three things that must be present for burning to take place.
- I can explain what happens if the amount of oxygen is reduced or runs out, or if fuel is removed.

EXPLORE:

Explain to children that they are going to find out more about another non-reversible change which also involves air/oxygen during this lesson: burning. Ask them to give examples of things that burn.

Ask: *What do you think happens when something burns? What do we need to start it burning? Why does something stop burning?*

Explain that because candles are dangerous, can cause burns and set fire to things if mishandled, you, the teacher, are going to light the candles and they are going to make careful observations of what they see.

Provide children with paper and pencils to draw observations and make notes. A digital camera would be useful to record additional evidence of the changes taking place. Challenge children, working together in pairs, to come up with as many observations as possible using scientific vocabulary to describe the changes that they see taking place. Suggest they draw a sketch of the candle and annotate it to show their observations. How many changes do they notice? Allow children to observe.

Ask: *When the candle was first lit, what did you notice? What colours can you see in the flame? Do they change? Does the flame size stay the same or does it change? What do you think the wick does? What do you think the wax does? What is burning – the wick or the wax? What do you notice after the candle has burned for some time? Where has the wax gone? What do you think is happening to the wax?*

Allow children to complete their observations.

Ask: *What do you think is needed for something to burn?*

Establish that three things are needed: fuel, oxygen and heat. In a burning candle, where are those things? Children should be able to identify the fuel as the wax, the oxygen in the air, and the heat as the flame.

ENQUIRE:

Explain to children that they are going to investigate what happens if oxygen cannot get to the candle.

Ask: *What do you think will happen if a jar is placed over the candle? Will it make a difference if the jar is a large jar or a small jar?*

Demonstrate using a small and a large jar. Measure the time it takes for the flame to be extinguished in each jar.

Ask: *Why did the candle go out more quickly in the smaller jar?*

Challenge 1 helps children to consolidate the ideas about the relationship between burning and oxygen that they have begun to develop in the Explore activity. In Challenge 2 children with a more developed understanding investigate the relationship further. In Challenge 3 children who can clearly explain what is needed for burning to occur apply their knowledge to the problem of bush fires in Australia.

Challenge 1: Children record the time taken for candles to stop burning in different sized jars, then describe the relationship they observe between the amount of oxygen in the jar and how long the candle stays alight

Provide the children with a Recording grid (Resource sheet 1). Ask them to record the results for Jar 1 (small jar) and Jar 3 (larger jar) and then estimate what the results would be for Jar 2 and Jar 4. They should describe the relationship they have observed between the amount of oxygen in the jar and how long the candle stays alight.

Challenge 2: Children measure the capacity of jars of different shapes and sizes and predict how long candles placed under them will stay alight

Ask them to explain their predictions in terms of the relationship they observe between the amount of oxygen in the jar and how long the candle stays alight. Show the children how to measure the capacity of the two jars used in the test by pouring water from a measuring jug into the jars and measuring in millilitres.

Ask: *How many millilitres of air were in each beaker? What difference did it make to the length of time the candle stayed alight? Can you find some other containers and estimate whether the candle would take more or less time to go out?*

Encourage the children to test containers of different shapes as well as sizes. They should log their results in a chart, together with the time they predict it will take for each candle to go out. If necessary, show them a possible layout for their Results chart (Resource sheet 2). They should explain their predictions with reference to the relationship they have observed between the amount of oxygen in Jars 1 and 3 and how long the candle stays alight.

Challenge 3: Children apply their understanding of the three things needed for something to burn to investigate the problems of large-scale bush fires in Australia

Ask: *What does a flame need in order to burn? What do we know already about putting out fires?*

The children may refer to fire safety demonstrations they have seen or practical examples like dousing the flames of a bonfire. They can use the internet to find out about recent bush fires in Australia.

Ask: *Why are bush fires such a risk in Australia? How do they start? What causes them to spread? How big an area do they cover? How long do they burn for? What is done to put them out? What can be done to prevent them?*

Prompt them to remember to check for mention of the three things – fuel, oxygen and heat – that must be present if a flame is to burn. Ask them to produce a fact sheet about bush fires in Australia.

Key information:

Oxygen only makes up about 20% of air.

REFLECT AND REVIEW:

Encourage children who completed Challenges 1 and 2 to share conclusions about the relationship between the amount of oxygen and burning time. Ask children that completed Challenge 3 to share what they have found out about bush fires and how they are fought and prevented. Prompt them, if necessary, to explain clearly how the three things that fires need are reduced or removed.

Provide large Odd one out cards (Resource sheet 3) for children to discuss in groups. Remind children how to play Odd one out. Remind them that there are no 'right' answers, because it depends on the criteria they choose; for example, cheese, sugar and salt. Cheese might be odd because it does not dissolve, but the others do; sugar could be odd because you might put cheese and salt in an omelette, but not sugar. When they are happy with an odd one out they have chosen, they should share it with another pair, swap cards, and begin the process again. Allow several swaps to take place.

EVIDENCE OF LEARNING:

Listen to children's responses throughout this lesson. Do they observe the changes involved in burning a candle carefully, using appropriate vocabulary to describe what they see? Can they identify examples of the reversible and non-reversible changes involved in a candle burning? Can they name the three things that must be present if burning is to take place? Can they describe the relationship between the amount of available oxygen and the length of time something burns? Are they able to explain what happens if the amount of oxygen is reduced or runs out, or if the fuel supply is removed?

MATERIALS: ALL CHANGE!

LESSON 5: HOW LONG DOES IT TAKE FOR THINGS TO RUST?

LESSON SUMMARY:

This is the second of two lessons exploring a non-reversible change: rusting. By the end of this lesson children will have collated the results of the observation enquiries begun a couple of weeks before in Lesson 3, drawn conclusions and presented them to their peers. They will also have applied their findings to a real life situation.

Key vocabulary:

material, compare, contrast, solid, liquid, gas, rust, oxidise, oxygen, corrode, tarnish; types of metal: iron, steel, chromium, tin, zinc; reversible, non-reversible, change of state

Resources:

The beakers containing iron and other materials that were being observed in different conditions to investigate rusting; the children's observation records

Health and safety:

Metal paint is not removed by washing with soap and water. Take great care to avoid spills and keep quantities to a minimum, as only a small amount is required to paint a nail or two.

Key information:

Almost all metals corrode to some extent when they come into contact with air over a long period of time. This results from a chemical process called oxidation. If the metal contains iron, the iron reacts with oxygen in the presence of water to form a reddish coloured solid called rust. The rate of rusting can be slowed down by treating the iron in different ways, including painting.

National curriculum links:

Explain that some changes result in the formation of new materials and that this kind of change is not usually reversible, including changes associated with burning and the action of acid on bicarbonate of soda

Learning intention:

To present the findings and conclusions from investigations into rusting

Scientific enquiry type:

Observing changes over different periods of time

Working scientifically links:

Reporting and presenting findings from enquiries, including conclusions, causal relationships and explanations of and degree of trust in results, in oral and written forms such as displays and other presentations

Success criteria:

- I can recognise that metals corrode; for iron and steel this is called rusting.
- I can describe the process of rusting and identify it as a non-reversible change (requiring water and air).
- I can explain what conditions are needed for something to rust.
- I can interpret the results of my observations over time and draw conclusions from them.
- I can present my conclusions and relate them to a real life context.

EXPLORE:

Revisit the problem of Lucy's birthday bike (Lesson 3, Slideshow 1). Remind children of the question she asked: How can I keep my bike looking as good as new? and discuss the different responses she received.

Ask children to think about these ideas again and what advice they might give Lucy now that they have done their investigations. Children do not need to respond at this point in the lesson because they will do so as part of their challenge, but they should think about the evidence their observations over time will provide to support or refute the statements in the image.

ENQUIRE:

Explain to children that during this lesson they will be looking at the results they have collected from their investigations of rusting and deciding what their conclusions are.

Children work in their original challenge groups. The challenges vary according to the level of detail required in reporting and the level of support provided.

Challenge 1: Children prepare a poster showing the results of their investigation into how much iron nails rust in different conditions

Ask: *What do you need to include in your poster to tell someone what your question was? What type of enquiry was it? What were you comparing? What observations did you make and how often? What did you find out? Why should someone trust your findings?*

Challenge 2: Children prepare a short PowerPoint presentation (5 slides maximum) reporting the results of their investigation into which method is most effective at slowing down (or stopping) the rusting process, illustrated with examples of where different ways of protecting structures against rust are used

Ask: *What do you need to include in your presentation to tell someone what your question was? What type of enquiry was it? What were you comparing? How did you collect your data? What are your conclusions? Can you find evidence of the best methods for preventing rust that your findings showed used in real life situations?*

Challenge 3: Children prepare a short PowerPoint presentation (5 slides maximum) reporting the results of their investigation into which metals and other materials rust, and the application of this knowledge to everyday objects

Ask: *What do you need to include in your presentation to tell someone what your question was? What type of enquiry was it? What were you comparing? How did you collect your data? What are your conclusions? Can you find evidence of the best methods for preventing rust that your findings showed used in real life situations? What can you find out about galvanising? What is a sacrificial metal? Would a plastic car be a good idea?*

REFLECT AND REVIEW:

Ask children from different groups to share their posters/PowerPoint presentations with the rest of the class. Encourage the audience to ask questions for information, for clarification and for opinion.

After listening to the different groups, return again to Lucy's problem and ask children what advice they would give her.

EVIDENCE OF LEARNING:

Listen to children's responses throughout the lesson, taking particular note of what they included in their posters/PowerPoint presentations. Are they able to describe the process of rusting, noticing the changes that take place? Do they identify rusting as a non-reversible change that requires water and air to take place? Do they recognise the variables that affect rusting? Can they describe the relationship between the different conditions they observed and the amount of rusting? Did they make sufficiently detailed observations, commenting on how the amount of rusting and the time to rust varies in their own investigation? Were they able to present their results clearly, justify their conclusions and answer questions effectively?

MATERIALS: ALL CHANGE!

 ## ENRICHMENT LESSON 1: WHAT WOULD MAKE THE BEST ROCKET FUEL?

LESSON SUMMARY:

In this lesson children use knowledge gained from Lesson 2 to investigate a non-reversible change that takes place when an effervescent vitamin C tablet and water are combined. By the end of this lesson they will have worked in a group to identify a question and plan and carry out a fair test, and will be able to explain that the changes they observed are non-reversible.

Key vocabulary:

material, mixture, solid, liquid, gas, powder, particle, tablet, bubbles, carbon dioxide, reversible, non-reversible, change, reaction

Resources:

Narrow measuring cylinders or small beakers, water, effervescent vitamin C tablets (one per group), small containers with snap-on lids. Stomp rocket (optional), sticky tack, water in jugs

Health and safety:

The Stomp rockets in this investigation can go off at a considerable force. Warn children not to return to look at a rocket if it seems to be taking a long time to blast off. Consider use of safety glasses, if available (see Be Safe! section 13).

Key information:

When an effervescent vitamin C tablet is dropped into water, a chemical reaction takes place. The reaction produces a lot of carbon dioxide gas very rapidly, suggesting it is a chemical reaction rather than a change of state. It is a non-reversible change.

National curriculum links:

Explain that some changes result in the formation of new materials and that this kind of change is not usually reversible, including changes associated with burning and the action of acid on bicarbonate of soda

Learning intention:

To plan and carry out a fair test enquiry to investigate a non-reversible reaction that produces carbon dioxide

Scientific enquiry type:

Comparative and fair tests

Working scientifically links:

Planning different kinds of scientific enquiries to answer questions, including recognising and controlling variables where necessary

Success criteria:

- I can describe the process that occurs when a fizzy tablet is added to water.
- I can explain why the change is non-reversible.
- I can work with my group to plan and carry out a fair test to investigate what makes a fizzy tablet rocket fly highest.

EXPLORE:

Provide groups of three children with one effervescent tablet and beakers or measuring cylinders containing a small amount of water. Ask them to break off a small piece of tablet, add it to the water and observe what happens. Build up to the full tablet.

Ask: *What do you notice happens? How rapidly is the gas produced? Where does the gas come from? Was it in the water before the test? Has it come from the tablet somehow?*

Give children a set of Concept sentences (per group) (Resource sheet 1) and ask them to use the words to create a sentence describing the process that has taken place.

ENQUIRE:

Explain to children that today they are going to investigate what is the best mix of rocket fuel for a rocket.

First, show children how to use a small container with a lid to make a simple rocket that they will be able to adapt later.

The container and lid need to be as dry as possible and a snug fit. Put a small amount of water in the bottom of the container and stick a blob of sticky tack inside the lid. Push the tablet into the sticky tack, to keep it away from the water until you are ready to set off the rocket. Quickly turn the container upside down and place it on its lid in a large shallow bowl. Warn children not to return to their rocket before it has taken off!

Ask: *What might make a difference to how high the pocket rocket flies?*

Collect children's ideas and note them on a flip chart or board under the heading: Variables we could change. Prompt them, if necessary, with three example questions: Does the size of container make a difference? Does the amount of solid make a difference? Does the type of liquid make a difference?

Remind them that they have only limited resources available: water, effervescent tablets, different sized containers and sticky tack.

Children should work in small groups to choose a variable to test and plan a fair test, predicting when the rocket will fly highest. Remind children of the success criteria for a fair test. Children carry out their fair test, evaluate the prediction they made and identify any further testing that might be necessary.

The challenge is the same for all children: to work in a group to plan a fair test to find which is the best fuel for their rocket. Share the fair test generic success criteria with all the children and provide a fair test planner and blank tables and bar charts. Some groups will need more support to identify variables to change, measure and control; to decide how to collect and present data; and to interpret and communicate their findings. Explain to the children that this is an opportunity for everyone to challenge themselves to work as independently as possible, plan their test efficiently and collect accurate data.

Ask: *What question is your group investigating? What is it you are trying to find out? What is your prediction? When do you think the rocket will fly the highest? Why do you think that? Which variables are you going to change and keep the same? What will you measure? How will you record your results? How will you tell others what you have found out?*

Key information:

In a Stomp rocket, or similar, pressure increases because of the physical action of pumping. The amount of gas (air) in the rocket increases as a result.

REFLECT AND REVIEW:

Ask each group to complete the following newspaper headline:

Breaking news: Best Ever Rocket Fuel Invented – just mix together …

Either take children outside to show them a Stomp rocket or show them a video clip of a Stomp rocket (Video 1).

Ask: *What is the difference between the rocket shown in the video and the rockets they have made? Why does the Stomp rocket launch into the air?*

EVIDENCE OF LEARNING:

Listen to children's responses throughout the lesson. Are they able to describe what happens when an effervescent tablet is added to water? Are they correct in their use of language, avoiding incorrect use of vocabulary such as 'dissolve' or 'melt'? Can they explain effectively that the change that takes place is non-reversible? Did they work well with their group as they completed their fair test? Did they identify difficulties with measuring and suggest how the accuracy of this could be improved? Can they compare different kinds of rockets and explain the differences?

MATERIALS: ALL CHANGE!

ENRICHMENT LESSON 2: WHAT ARE THE BUBBLES IN HONEYCOMB TOFFEE?

LESSON SUMMARY:

In this lesson children observe the process of making honeycomb toffee. By the end of this lesson they will have identified the changes that happen to the materials used in the recipe.

Preparation required: If possible, arrange for children to watch honeycomb toffee being made during the lesson.

Key vocabulary:

material, compare, contrast, solid, liquid, gas, powder, particle, dissolve, melt, boil, reversible, non-reversible

Resources:

For the honeycomb toffee: 100 g sugar, 2 tablespoons of syrup, 1 heaped teaspoon of bicarbonate of soda, 1 tablespoon of vegetable oil for oiling the pan, a large heavy-based saucepan, a wooden spoon, a tin lined with aluminium foil (to save on washing up!) and a means of heating, such as a portable stove or access to the school kitchen. Challenge 3 – honeycomb toffee that has cooled, samples of pumice stone, sponge cake, bread, expanded polystyrene, hand lenses, access to sources of information about these materials

Health and safety:

Adult supervision is required when making honeycomb toffee (see Be Safe! section 16).

Key information:

The bicarbonate of soda reacts as it heats up and releases carbon dioxide gas. The gas causes the toffee mixture to froth and bubble, and the gas becomes trapped in the sugar/syrup mixture.

National curriculum links:

Explain that some changes result in the formation of new materials and that this kind of change is not usually reversible, including changes associated with burning and the action of acid on bicarbonate of soda

Learning intention:

To describe the changes of state that take place as honeycomb toffee is made and compare it with other materials containing bubbles

Scientific enquiry type:

Observation over time

Working scientifically links:

Reporting and presenting findings from enquiries, including conclusions, causal relationships and explanations of and degree of trust in results, in oral and written forms such as displays and other presentations

Success criteria:

- I can observe carefully how the mixture changes as honeycomb toffee is made.
- I can identify the reversible and non-reversible changes that take place.

EXPLORE:

Show children the video of Honeycomb toffee being made (Video 1).

Ask: *What changes did you observe as the toffee was made? Were the changes reversible or non-reversible?*

Explain that because making honeycomb toffee involves heating up materials to high temperatures, you are going to show them how it is done. They are going to make careful observations, thinking about the processes that are taking place, and maybe taste some of the toffee when it cools later!

Demonstrate making the honeycomb toffee using the Honeycomb toffee recipe (Resource sheet 1).

ENQUIRE:

Explain to children that the challenge is either to sequence the changes they observed using the correct scientific vocabulary or to apply their understanding of the way the bubbles are produced in honeycomb toffee to a comparison of other materials that have holes. The challenges are differentiated by the level of detail required in the explanations (moving from describing, to explaining, to analysing). Group children according to how secure their understanding is of the changes they observe when the toffee is cooking.

Challenge 1: Children create concept sentences to describe the processes involved in making honeycomb toffee

Provide the children with Key vocabulary cards (Resource sheet 2) to use in making concept sentences. Ask them to describe each stage of the process of making honeycomb toffee, from start to finish. Their sentences should include all the words on the cards, plus any extra they need to add.

When they have completed their sentences, encourage the children to check them.

Ask: *Have you described all the processes that are involved? What happened to the sugar? How did it change? What happened to the bicarbonate of soda when it was added? Was it a reversible change or a non-reversible change? Do you need to add any more scientific words?*

Challenge 2: Children create a flow diagram to communicate the processes involved in making honeycomb toffee

Ask the children to sequence the process of making honeycomb toffee into a series of steps. Remind them of the stages they observed.

Ask: *What happened to the sugar and syrup? What changes did they go through? Why did that happen?*

Prompt them, if necessary, to link each stage in the process with arrows (use the Life cycle flow diagram, Resource sheet 3, as an example) and to use key vocabulary in their labels.

Challenge 3: Children use a Compare and contrast grid to make comparisons of two apparently similar materials

Ask: *What did you notice about the structure of the honeycomb toffee? Can you think of any other materials that appear similar?*

The children might suggest polystyrene foam, natural sponges, sponge cakes, bread and pumice stone. How are the 'holes' in the material caused in these different cases?

The children should use hand lenses to look closely at the structure of each material and choose one to compare systematically with the honeycomb toffee. Ask them to complete the Compare and contrast grid (Resource sheet 4), answering the question prompts as fully as possible. They will need to use other sources of information to answer some of the questions.

<table>
<tr><td>

Key information:

The examples used all involve the production of gases in different ways. Some are mainly physical (pumice stone), some chemical (baking powder in cake) and some biological (yeast in bread-making).

</td></tr>
</table>

REFLECT AND REVIEW:

Show the whole class the samples of materials Challenge 3 children compared and researched during the lesson. Provide samples for each group to look at and discuss, and send an 'envoy' Challenge 3 child to each group to provide some research information about the wider range of materials, and how and why the bubbles were formed.

Ask: *What did you notice? How were they similar and different? What size were the bubbles? Which material had the biggest bubbles? And the smallest bubbles? How were the bubbles made? Which material is the odd one out and why?*

EVIDENCE OF LEARNING:

Listen to children's responses throughout this lesson. Do they observe carefully, describing how the mixture changes as the honeycomb toffee is made? Are they able to describe changes as reversible and non-reversible, using key vocabulary with accuracy? Can they compare and contrast materials that appear to have similar properties? Can they find out more about how they were made?

FEEL THE FORCE

INTRODUCTION

In Year 3 children learned about how contact and non-contact forces make things start and stop moving. This module builds on these ideas and develops an understanding of how forces including gravitational attraction and drag forces – friction, air resistance, water resistance, and upthrust in water – affect movement. Children learn how mechanisms including levers, pulleys and gears allow a smaller force to have a greater effect, and they use this knowledge in different investigations. When working scientifically, children plan and carry out fair test and pattern-seeking investigations, observe carefully, record accurate measurements, and construct different mechanisms. They look at scientific ideas from the past and carry out an activity to find evidence to support or refute famous scientists' ideas. They make predictions as a result of carrying out simple activities and go on to plan new investigations. There are opportunities to develop graphing skills as well as communication and presentation skills.

Key vocabulary:

air resistance, Aristotle, balanced , balanced forces, bevel gears, clockwork, cogs, compress, extend, effort, force arm, forces, force, friction, force arrow, fulcrum, gravity, Galileo, gear ratio, gears, gear trains, lever, lift, machine, mechanisms, movement, Newton, Newton meter, pinion, pivot, pulley, pull, push, rack, resistance, rotary motion, simple machines, speed, time, unbalanced force, upthrust, water resistance, weight arm, wheel

FACT FILE:

Forces are at work on everyday things all the time. Everything that changes speed, stops, starts and changes direction has forces acting on it. These forces are invisible and only their effects are noticed, which is why this area of the curriculum is challenging, as the way forces work is often counter intuitive. The simple definition of a force is that it is as a result of a push or a pull, so gravitational attraction is a pulling force – a force that works between bodies at a distance. Another definition of force is the capacity to do work or to cause something to change.

There are two types of forces – those that work at distance and those that are in contact. Gravity and magnetism work at a distance, whereas friction, air resistance and water resistance work in contact. If an object is stationary or moving at a constant speed, then the forces acting on it are balanced. Newton's first law says that an object will stay still or, if moving, will continue to move at the same speed and in the same direction unless it is acted on by a force. Unbalanced forces cause changes to movement (start, stop, speed up, slow down and changes of direction).

Forces are vector quantities, which means that they have both size and direction. The use of force arrows will help children to see which force is acting, how big it is and in which direction it acts. Gravity is the force that pulls things towards the centre of the Earth.

Scientifically, mass is used to describe the amount of matter in a body and is measured in grams and kilograms. The universally accepted measurement of force is Newtons. Weight is a downward force due to the pull of gravity and so is measured in Newtons, which are named after Sir Isaac Newton, who created laws of motion. On Earth it is possible to use the units of mass and weight interchangeably, as we often refer to weight when we mean mass. However, on the Moon your weight would be less because the gravitational pull on the Moon is less, but your mass, the amount of matter in your body, would stay the same. There is no single force of gravity. The heavier the object, the greater the gravitational force exerted on it. Aristotle believed that the heavier the object, the faster it falls. It was Galileo who suggested that it was not the mass that affected the time taken to fall. The heavier object needs a greater pull of gravity to get it moving at the same speed as a lighter object, which needs less force to make it move. The tendency of all objects to keep moving at the same speed and in the same direction is called inertia.

Friction is the force that exists between the surfaces of two objects that are in contact with each other, when at least one of them is moving. If a block is pushed across the table, the initial movement, the push, starts the block moving. It stops because the force of friction works against this movement. If the surface is rough it has greater frictional force. Lubricants are used to counteract the effect of friction between two moving surfaces.

Air resistance is the force that opposes the movement of objects in air. When objects move through air, the air pushes against them and slows them down. Objects with greater surface area create more air resistance because they have to push more air out of the way. Unlike most other forces the effect of air resistance increases as the speed of the object increases, because faster objects crash through the air more violently. When things fall through the air the weight does not affect the time taken to fall if the surface area is the same. Galileo identified that if there was no air resistance then all things would hit the ground at the same time. He might have tested his ideas by dropping balls from the leaning tower of Pisa and finding that objects of the same size but different masses fell at the same rate. To make things move, even in the air, objects have to overcome inertia. So, a heavier object needs a bigger force to keep it moving at the same speed.

Water resistance occurs when an object moves through water or across the top of water, pushing against it and slowing it down. Water resistance is independent of the speed of the object. The more molecules that are in a liquid, the greater the effect of the resistance. The factor that affects the movement most is the surface area of the object in water.

The effects of both water and air resistance can be overcome by streamlining to reduce the surface area in contact with the air or water.

Water resistance and air resistance are both forms of friction and are used to explain how moving objects slow down. They should not be confused with upthrust. Upthrust determines whether something floats or sinks in air or water, due to the object's weight.

If a ball of clay, which is relatively heavy but has a small surface area, is dropped in water, the upthrust pushing up on the clay ball is unable to balance the downward force of gravity and the ball sinks. If an object is heavy for its size, like the clay ball, there is not enough upthrust to support it and make it float. For an object to float, the weight of water it displaces must be at least equal to the object's weight, so that the two forces of upthrust and gravity are balanced. Therefore, if the clay is reshaped to form a larger bowl shape, it displaces more water and floats.

Levers, pulleys and gears are simple machines, or mechanisms. They are devices that make things easier to do, whether it is lifting, turning or changing the direction of movement. A lever is a simple machine that makes a small force have a greater effect; it involves moving a load, the thing you want to lift, around a stationary part of the lever, called the fulcrum. There are a number of different types of levers, depending on where the force is applied. The force is called the effort. There are three types of levers: class 1 are seesaw types; class 2 are like a wheelbarrow; and class 3 are like a pair of tweezers. In this module, only class 1 levers are used. Gears are wheels with teeth called cogs, which mesh together so that when one is turned the other moves as well. If the size of the gears (and therefore the number of cogs) are changed, then the movement of the smaller gear can be increased. Gears also change the direction of movement. If one gear is turned clockwise then the adjoining gear turns anticlockwise. Clocks, tin openers and bicycle gears are all examples of the use of different sized gears which change the amount of movement. A pulley is a simple machine consisting of a wheel with a rope running over the top of it. One end of the rope is fixed to the object that needs to be lifted and the other end is pulled by a person or machine to raise the object. Pulleys work by allowing the movement to be a pull down rather than a lift up, therefore getting gravity to help when the heavy object is lifted. The more pulleys that are used to lift objects, the less pulling force needed.

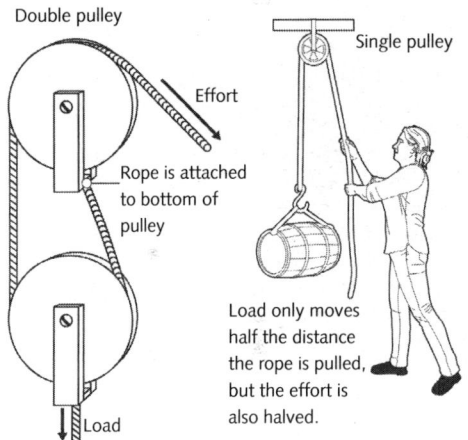

Double pulley

Single pulley

Effort

Rope is attached to bottom of pulley

Load

Load only moves half the distance the rope is pulled, but the effort is also halved.

Common misconceptions:

- Children use the everyday understanding of the term 'force' – that is, when someone makes you do something you do not want to do. It is also related to the Armed Forces, in certain schools.
- Children identify motion as moving or not moving.
- Children also think that movement stops when things 'run out of push' rather than because there are other forces acting on them. They may think that to keep an object moving you need to keep giving the object the force (push). This common misconception is because of the invisibility of the other forces at work. To help overcome this idea the use of arrows to define size and direction of the forces is written into the module. This is particularly important because force is a vector and therefore should have both its size and its direction identified.
- Children believe that a stationary object has no forces acting on it whereas the reason the object is stationary is because the forces acting on it are balanced.
- Children and many adults believe that heavy things fall faster than light objects, when it is the surface area and air resistance that affect the way objects fall.

FEEL THE FORCE

Key vocabulary:

mass, gravity, Newton meter, friction, smooth, rough, movement

Resources:

Newton meters (2.5 N, 5 N, 10 N, and 20 N), modelling clay, toy vehicles, three sizes of matchbox, some of which have different materials glued to the base (for example, rough sandpaper, aluminium foil, rectangular sections cut from rubber gloves or thin foam rubber, cotton cloth)

Key information:

There are different types of forces. Some work in contact with objects, such as friction, air resistance and water resistance; others work at a distance (non-contact forces), such as magnetism and gravity. Friction is a force that opposes motion between moving surfaces in contact. The size of this force depends on the properties of the surfaces (for example, roughness).

Key information:

Newton meters have two scales, one in grams and one in Newtons. The gram is the standard measurement of mass. The Newton (named after Sir Isaac Newton, who created laws of motion) is the unit for measuring force and weight. Ask one child in a group to hold the Newton meter. A second child loops a finger around the hook of the Newton meter and to tries to pull to the number of Newtons suggested by a third child. The child who is pulling should look at the scales as they do so. Let children swap roles and repeat until all three have pulled the Newton meter.

LESSON 1: HOW CAN WE MEASURE FORCES?

LESSON SUMMARY:

This lesson builds on children's Year 3 work, where they observed that forces exist everywhere – both contact and non-contact forces, and that these forces act on all things to make them stop and start moving, speed up, slow down and change shape. In this lesson children extend their understanding of friction by learning how to measure forces using a Newton meter. By the end of the lesson they are able to use a Newton meter accurately.

Preparation required: Cut out and laminate six sets of cards for the Forces card sort (Resource sheet 1). Make a small hole in one end of the matchbox trays to insert a Newton meter hook (see below). Stick the four suggested materials to the bases of the three different sizes of matchbox. Leave some matchboxes unchanged, as controls.

National curriculum links:

Identify the effects of air resistance, water resistance and friction, which act between moving surfaces

Learning intention:

To measure, using appropriate units, friction between moving surfaces as part of an investigation into how the surface area and materials affect friction

Scientific enquiry type:

Noticing patterns

Working scientifically links:

Taking measurements, using a range of scientific equipment, with increasing accuracy and precision, including taking repeat readings when appropriate

Success criteria:

• I can use a Newton meter with accuracy.
• I can record my results in a table.
• I can identify the effect of friction in my investigations.
• I can explain my results using scientific vocabulary, including the word 'friction'.

EXPLORE:

Organise children into groups of three. Give each group a set of Forces cards (Resource sheet 1). Ask them to decide whether the statements on the cards are true or false. This reminds them of Year 3 work on forces and indicates what they know about friction and the measurement of forces. After the card sort ask children which decisions they are sure about and why, and which they need to investigate.

Give each group rectangles (approximately 10 x 5 cm) of rough sandpaper and metal foil. Encourage them to talk about how the materials feel.

Ask: *Which surface would be better for a playground slide? Why? Which will it be easier to drag a heavy box across? Why?*

Explain to children that there is a 'rubbing' force acting between the bottom of the box that is being dragged and the surface that the box is resting on. Introduce 'friction' for this force. Demonstrate friction by giving a toy vehicle a push to get it rolling along a table. Explain that the friction between the wheels and the table will slow down and stop the vehicle.

Show children Newton meters, for example, 2.5 N, 5 N, 10 N and 20 N. Give each group of four children two different Newton meters to compare.

Ask: *What is the same about the two Newton meters and what is different?*

Tell children to look at the size of the spring and of the scales, and ask them what they think the scales are measuring.

Repeat the activity, with the child pulling the Newton meter, closing their eyes, and saying when they think the reading on the meter is correct. The third child notes the actual reading. Let them swap roles.

ENQUIRE:

Show children a matchbox weighted with modelling clay. Ask which Newton meter they think would be right for measuring the force needed to move the matchbox across a table. Explain that they are going to use Newton meters to measure how the material and the area of two surfaces affect the friction between them.

Explain that they should use only the Newton scale and check it is set at zero before taking each reading. Advise them they need to weight each matchbox with 150 g of modelling clay for them to register on a Newton meter, which reads up to 2.5 N.

The challenges are differentiated by the amount of data children need to collect, present and interpret. Children work in groups of three. The challenges are presented on the Challenge slides to be displayed on the board, or printed out and placed in the centre of the table.

Challenge 1: Children measure the force needed to move different materials across a table

Give the children the largest matchboxes with different materials on their bases and another of the same size with nothing on its base as a control. Ask them to measure the amount of force in Newtons that is needed to move each matchbox.

Ask the children to record their results in the Matchbox force table (Resource sheet 2) and to calculate the average reading for each surface. When all matchboxes have been tested by each child, ask them to place the matchboxes in order of the force needed to make them start moving. Ask them to write three 'If…then…because…' sentences about the amount of force needed to move the matchboxes.

Challenge 2: Children measure the force needed to move medium sized matchboxes with different materials on their bases

Ask the children to carry out the same procedure as in Challenge 1, but using medium sized matchboxes with smaller base areas

Encourage them first to predict which matchboxes they think will require the most force and the least force to make them move and why.

Ask: *How will you record your results? How often should you repeat the test? Do you think three readings for each matchbox is enough?*

Explain to the children that they can record their results in any way they think is suitable and ask them to calculate the average reading for each matchbox surface.

After testing all the matchboxes, the children should put them in order of the amount of force needed to keep them moving and write three 'If…then…because…' sentences linking the material on the matchbox and the force needed to move it.

Challenge 3: Children measure the force needed to move the smallest matchboxes with different materials on them and draw a bar graph of results from the most to the least reading for friction

Ask the children to carry out the same procedure as in Challenge 1 with the smallest matchboxes (smallest base area). Encourage them to predict which matchbox will require the greatest and the least force to start them moving, and to explain why. Ask the children to write the prediction as a ranking, and to think about how to record their results and how often to repeat the test.

Ask: *Do you think the three readings for each matchbox is enough?*

Ask the children to calculate an average reading for each matchbox surface and to present their data in a bar graph, using the generic graph template if required.

REFLECT AND REVIEW:

Arrange children in new groups of six, including two from each challenge group. (Or do this as a whole class using results from the three groups.) Ask them to make a new bar graph that compares the average results for different surface areas (Class results bar graph; Resource sheet 3).

Explain that the materials stuck to the bases of the boxes in Challenge 1 were twice the surface area of the those in Challenge 2 and three times the surface area of those in Challenge 3. Ask children to look at the bar graph and discuss, in pairs, any relationship they find between the surface area and the force needed to move the box.

Ask: *Do the forces you measured match your predictions? Can you explain any that didn't?*

Encourage children to consider what might happen to the force readings if the amount of modelling clay was increased to 300 g. Demonstrate using two large matchboxes with material stuck on them. Discuss the results.

EVIDENCE OF LEARNING:

Watch and listen to children as they complete the challenges. Can they design a table to record their measurements? Do they use the Newton meter accurately? Do they repeat their readings and explain why they need to? Can they calculate averages? Can they transfer their results to a bar graph accurately? Do they use 'friction' in 'If…then…because…' sentences? Can they describe the relationship between the size of surface area and the force needed to keep the matchbox moving? Can they predict the impact of increasing the mass in terms of force needed?

Key information:

It is important for children to experience the amount of force in Newtons, so that they understand the measurements they take in this and subsequent lessons.

Key information:

If possible, make the surface area of the largest matchbox twice that of the medium one and the surface area of the medium sized matchboxes twice that of the smallest one (this will be relevant to the Reflect and review section of the lesson).

FEEL THE FORCE

LESSON 2: WHY DOES AN OBJECT FALL?

LESSON SUMMARY:

In this lesson children identify how scientific evidence is used to support and refute ideas, testing the explanations of Aristotle and Galileo about how things fall. Children investigate and find evidence for these ideas, exploring gravity as a non-contact force. By the end of the lesson they understand that gravity pulls objects towards the Earth and that an object's surface area, not its mass, affects the time it takes to fall.

Preparation required: Children are going to time accurately the fall of objects over a short distance, so they should work in the hall or anywhere where they can gain extra height by standing on PE benches to drop objects.

Key vocabulary:

gravity, falling, surface area, weight, mass, air resistance

Resources:

Objects to drop to demonstrate something falling, empty camera film canisters, modelling clay, good quality cupcake cases, timers

Health and safety:

Children should not stand on chairs. Remind them to drop objects safely. See section 13 of 'Be Safe!'.

National curriculum links:

Explain that unsupported objects fall towards the Earth because of the force of gravity acting between the Earth and the falling object, and identify the effects of air resistance, water resistance and friction, which act between moving surfaces

Learning intention:

To use evidence to explain how objects fall through the air

Scientific enquiry type:

Carrying out comparative and fair tests

Working scientifically links:

Identifying scientific evidence that has been used to support or refute ideas or arguments

Success criteria:

- I can plan a comparative test to investigate my ideas about how objects fall.
- I can identify how scientific evidence has been used in arguments.
- I can explain why objects fall using scientific vocabulary.

Key information:

Remind children of their work on magnetism, which is another example of a non-contact force.

Key information:

If a piece of paper is held vertically in the portrait position and is completely flat and has no creases, it is possible that when released it will drop straight down to the floor as there is virtually no air resistance affecting it.

EXPLORE:

Select objects that are obviously different in size and weight.

Ask: *What will happen if I let go of these objects?*

Drop the objects one at a time. Ask children to discuss, in pairs, what is happening. Select some responses to write on the board. Remind children that if something moves, a force must be involved.

Ask: *What force is making the object fall?*

If necessary explain that the force is gravity. Write it on the board. Explain that gravity is a pulling force that acts between objects, in this case between the objects and the Earth, and that gravity is a non-contact force.

Select two objects of obviously different size and weight, for example, a paper clip and a book.

Ask: *What will happen if I drop these at the same time?*

Select some responses to record on the board. Drop the objects several times. They will hit the ground at the same time. Encourage children to discuss this and write their comments. Were they surprised?

Ask: *What will happen if I drop a flat piece of A4 paper held horizontally at the same time as a piece of A4 paper screwed into a ball?*

Drop the pieces of paper at the same time, from the same height. Ask children to discuss, in pairs, what they observe. Explain that as well as gravity pulling downwards another force acts upwards. This force acts in the opposite direction to gravity. Introduce 'air resistance'. Write it on the board. Ask why it affects flat paper more than screwed up paper.

ENQUIRE:

Explain to children that their challenge is to test the ideas of two scientists from long ago, and to decide who was correct. Show the Aristotle and Galileo challenge slideshow (Slideshow 1). Keep this on the screen as children carry out the challenges.

Key information:

When objects fall, gravity pulls them towards the centre of the Earth. The speed of the descent is affected not by an object's mass, but by the opposing drag force – air resistance. Without air resistance any objects dropped simultaneously hit the ground simultaneously.

Explain to children that almost 2500 years ago in Greece, Aristotle thought that heavier objects fall faster than lighter objects. Galileo, four hundred years ago in Italy, thought that all objects fall at the same speed, no matter how heavy.

Organise children into threes according to their understanding of forces, demonstrated through responses in the Explore section (Lesson 1).The challenges are differentiated by difficulty of task and type of data that children collect.

The challenges are presented on the Challenge slides to be displayed on the board, or printed out and placed in the centre of the table.

Challenge 1: Children compare falling objects of different weights

Explain to the children that they are going to use film canisters containing modelling clay to investigate gravity. Give each group Aristotle vs Galileo (Resource sheet 1). Remind them that they are collecting evidence to test Aristotle's idea (heavy objects fall faster than lighter objects) against Galileo's (mass [weight] does not make a difference to how fast something falls). Ask them to record the results on the challenge sheet and say whether their evidence supports or refutes each scientist's idea.

Challenge 2: Children compare the effects of air resistance on paper

Give each group Aristotle vs Galileo (Resource sheet 2). Ask them to follow the instructions to test the effects of gravity on flat and scrunched up paper. Remind them to collect evidence to test Aristotle's idea (heavy objects fall faster than lighter objects) against Galileo's (mass [weight] does not make a difference to how fast something falls). Ask how many times they will carry out the test. Remind them to test their ideas systematically, dropping things from the same height.

Ask them to record their results and to use them to decide who was right, Galileo or Aristotle, and to justify their answer.

Challenge 3: Children measure the effect of mass on the time cupcake cases take to fall

Give each group of three Aristotle vs Galileo (Resource sheet 3). Remind them to collect evidence to test Aristotle's idea (heavy objects fall faster than lighter objects) against Galileo's (mass [weight] does not make a difference to how fast something falls). Ask how many times they will carry out the test. Remind them to test their ideas systematically, dropping things from the same height, and to check for anomalous results.

Ask the children to use their recorded data to decide who was right, Galileo or Aristotle, and to justify their answer.

REFLECT AND REVIEW:

Ask each group whether their evidence supports Galileo's or Aristotle's idea and to justify this.

Show children (Video 1) (a feather and hammer dropped on the Moon). Ensure they know that the Moon's gravity is much weaker than Earth's.

Ask: *Which scientist's idea is being tested? Does the evidence support or refute his idea?*

EVIDENCE OF LEARNING:

Listen and watch during the challenges. Do children take and record readings accurately? Can they use findings to refute or support scientific ideas? Do they use scientific vocabulary? Can they describe how gravity pulls things towards the Earth? Do they recognise that mass does not affect how fast something falls? Can they explain that air resistance slows falling objects and so the surface area of an object affects this?

FEEL THE FORCE

LESSON 3: WHAT MAKES THINGS MOVE?

LESSON SUMMARY:

In this lesson children investigate how forces make things change direction, speed up, slow down, start or stop moving and use force arrows to represent these. By the end of the lesson they are able to draw force diagrams to show how blowing on bubbles affects their movement.

National curriculum links:

Explain that unsupported objects fall towards the Earth because of the force of gravity acting between the Earth and the falling object

Learning intention:

To use arrows to represent forces that make objects move in different directions

Scientific enquiry type:

Carrying out simple comparative and fair tests

Working scientifically links:

Planning different types of scientific enquiries to answer questions, including recognising and controlling variables where necessary

Success criteria:

- I can explain how to change the speed and direction of bubbles as they fall.
- I can use arrows to represent forces.

EXPLORE:

Put the toy vehicle on a table. Ask children what forces, if any, are acting on the vehicle. If necessary, explain that gravity pulls it downwards.

Ask: *If this is correct, why isn't the vehicle moving downwards?*

Explain that the table pushes upwards on the vehicle, providing an equal and opposite force. Demonstrate this. Ask a child to put a hand on the side of an inflated balloon; you put a hand on the opposite side. The balloon is held between the two hands. Explain that the balloon must not move in any direction. Push gently on the balloon: the child has to push back to keep it still. Push a little harder; the child has to do the same. Ask the child to explain what he could feel and what he/she had to do to keep the balloon still. Explain that whatever force you used, the child had to push back with the same force to keep the balloon still.

Repeat this with another child; this time place hands above and below the balloon. Push gently downwards to represent gravity; the child has to push upwards to represent the reaction force. Ask what had to be done to keep the balloon still.

Explain that we use arrows to show size and direction of forces. Hold a small arrow on one side of the balloon to represent your push. Ask the child pushing on the other side what size arrow to use. (If necessary, explain that the same size arrow is needed.) Push harder on the balloon and select a bigger arrow to represent the push. Ask what size arrow is needed to represent the opposing force.

Explain to children that the balloon represents the toy on the table. Ask which two forces are acting on it and how to show them using arrows. Show how to use arrows to represent forces on the stationary vehicle – one arrow pointing down represents gravity and another, the same size, pointing up represents the reaction force from the table.

Ask children what will happen if they push the vehicle.

Ask: *Is this another force? How could we show this with arrows? Will we need a bigger arrow?*

To move the vehicle a push force (a contact force) is needed; this can be represented by a different sized arrow.

ENQUIRE:

Explain to children that they are going to use arrows to show how they make objects move and change speed and direction. Blow some bubbles. Ask children to describe how the bubbles are formed. If necessary, explain to them that air pushes them.

Ask: *What will happen if I blow bubbles upwards? What force pulls them down? Does the same force act on the vehicle on the table?*

Explain to children that they are going to investigate how to use a force to make bubbles defy gravity. The three challenges involve the same activity, but are differentiated by how children report

Key vocabulary:

gravity, air resistance, friction, fast, slow, start, stop, change direction, fall, rotate, contact force, non-contact force, reaction force, balanced

Resources:

Bubble mixture, big toy vehicle, balloon, straws, fans (battery and paper), hairdryer, table tennis ball, laminated card arrows of different sizes

Key information:

Children will say that the table is in the way, stopping the vehicle moving downwards. They may not understand that two equal forces act on it: the table provides an upwards force, equal and opposite to gravity – this is a reaction force.

Key information:

It is possible to change bubbles' downward movement using another force. Bubbles are good for demonstrating 'invisible' forces because it is easy to change their speed and direction of movement by blowing the invisible air.

it. Group [children] ~~~~~~~~~~~~~~~~~~~~~~~~~~~~~~ e section and understanding of gravity and force ~~~~~~~~~~~

The challenges are presented on ~~~~~~~~~~~~~~~~~~~~~~~~~~~~~ on the board, or printed out and placed ~~~~~~~~~~~~~~~~~~~~~~~~~~

Challenge 1: Children investigate ~~~~~~~~~~~~~~~~~~~~~~ and direction of the movement of ~~~~~~~~~~~

Give each group a ~~~~~~~~~~~~ bble mixture, straw~~, and battery powered and paper fans for moving bubbles. A~~~~~~~~~~~ bubbles rotate as well as move upwards and sideways. Ask them to keep th~ bubbles ~~~~~ air for as long as possible.

Ask them to watch bubbles fall to the ground to see if they fall at the same speed. Ask them to draw labelled diagrams, including different sized force arrows, to show what happens when they blow on a bubble.

Challenge 2: Children investigate different ways to change the speed and direction of the movement of bubbles

Give each group a container of bubble mixture, straws, and battery powered and paper fans to move the bubbles. Ask them to make bubbles rotate as well as move upwards and sideways. Ask them to keep the bubbles in the air for as long as possible.

Ask children to watch bubbles fall to the ground to see if they fall at the same speed. Ask them to draw labelled diagrams, including different sized force arrows, to show what happens when they blow on a bubble.

Encourage them to write two 'If…then…because…' statements about keeping bubbles in the air.

Challenge 3: Children investigate different ways to change the speed and direction of the movement of bubbles

Give each group a container of bubble mixture, straws, and battery powered and paper fans to move the bubbles. Ask them to make bubbles rotate as well as move upwards and sideways. Ask them to keep the bubbles up in the air for as long as possible.

Ask children to watch as bubbles fall to the ground to see if they fall at the same speed.

Ask: *Can you name the forces acting on the bubbles?*

Ask children to draw labelled diagrams, including different sized force arrows, to show what happens when they blow on a bubble.

Encourage them to write three 'If…then…because…' statements about keeping bubbles in the air.

REFLECT AND REVIEW:

Share and discuss diagrams and statements from each group. Ask what would happen if they could make a bubble hover in the air without it moving up, down or sideways.

Ask: *What would the size of the force arrows be?*

Demonstrate by placing a table tennis ball above a hairdryer blowing upwards. It is possible to keep the ball in the air. Ask children to select force arrows to show what is happening. Remove the upward force from air – gravity pulls the ball to the floor.

EVIDENCE OF LEARNING:

Listen and watch children completing challenges. Do they use phrases such as changing direction, speed up, slow down and fall to describe motion? Can they recognise that moving air creates a contact force (a push) on bubbles? Do they use different sized arrows to represent different sized forces? Do they use arrows to represent the force from blowing or from a fan, the gravitational force? Do they recognise that gravity acts on everything? Can they identify forces acting on a stationary object? Can they explain that an object moves because forces acting on it are unbalanced?

FEEL THE FORCE

Key vocabulary:

air resistance, force, gravity, surface area, mass, weight, pull, measurement, test, variables, time, fall

Resources:

Different types of string, scissors, plastic bin liners, different sized small plastic, a range of materials, including tissue paper, plastic, fabric, card, paper

Key information:

When investigating air resistance it is important to encourage children to observe and record how their objects fall, not just the time it takes for them to fall. The reason for this is that some materials lack rigidity and as a result may float because they have a greater surface area. If a small hole is cut in the middle of the parachute this will allow air to slowly pass through the hole rather than spilling out over one side. The hole should also help the parachute fall in a straight line.

LESSON 4: HOW CAN WE SLOW DOWN FALLING OBJECTS?

LESSON SUMMARY:

In this lesson children plan and carry out a fair test investigation into air resistance, using parachutes. They make parachutes and measure the time taken for the parachute to fall. They will then use the results of their initial investigations to predict how they could improve their parachutes and plan and test out their ideas. By the end of the lesson children have identified which variables they could change to keep the parachute in the air for longer and identified the forces that act on a parachute.

National curriculum links:

Explain that unsupported objects fall towards the Earth because of the force of gravity acting between the Earth and the falling object, and identify the effects of air resistance, water resistance and friction, which act between moving surfaces

Learning intention:

To use test results about air resistance as a starting point for further investigative work

Scientific enquiry type:

Carrying out simple comparative and fair tests

Working scientifically links:

Using test results to make predictions to set up further comparative and fair tests

Success criteria:

- I can use the term 'gravity' in an explanation of why things fall.
- I can make a prediction.
- I can use this prediction to plan a fair test.

EXPLORE:

Organise children into groups of three and provide each group with the Parachute instruction card (Resource sheet 1) and the materials needed to make their parachute. Keep this activity to less than 15 minutes. Once children have all made their parachutes, ask them to test how long the parachutes take to fall to the floor from their head height. They should test their parachutes a number of times.

After they have done this, ask children what was the longest time recorded for a parachute to fall. Ask them if all the tests gave the same results, and if not, why they think there were variations in the time taken.

Encourage children to think of one change that they would make to the parachute to make it stay in the air longer. Ask them to think of a sentence beginning 'I think it would stay in the air longer if…' This helps children to focus their thinking not just on predicting the improvements, but also on giving reasons for the change/s that the improvements would result in.

ENQUIRE:

Ask children to identify the variables that they could change if they wanted to make a parachute that falls to the ground more slowly than their first test parachute. Ask them to choose one variable to test, identify the equipment they will need, and decide how they will record their results. They can use the generic fair test planner and blank tables if they choose.

All the groups have the opportunity to change different variables, for example, the size of parachute, the material it is made from, the weight of the toy, the length/type of string, the shape of parachute, etc. This flexibility means that children will all be investigating slightly different questions. The challenges are differentiated by the ways in which children are required to record and present their data, with increasing precision.

The challenges are presented on the Challenge slides to be displayed on the board, or printed out and placed in the centre of the table.

Challenge 1: Children compare how different improvements to their parachute affect the time it takes for the parachute to fall to the ground

Encourage the children to change a variable in at least three ways, for example, three different sizes (small, medium and large).

Ask: *Which variables will you keep the same (control)? How many times should you repeat each test? How will you compare and record the results?* (Children may choose to launch the parachutes at the same time and rank them according to the time taken to fall.)

When they have completed their tests, give the children some time to review their results.

Ask: *Which parachute took the longest to fall? What is the relationship between the variable you changed (the independent variable) and the time the parachute took to reach the ground? Can you explain this in terms of gravity and air resistance? Based on your results, what improvement would you make to your first parachute?*

Challenge 2: Children measure how changing an independent variable affects the time it takes for their parachute to fall to the ground

Encourage the children to change a variable in at least three ways, for example, by using three different types of material.

Ask: *Which variables will you keep the same (control)? How many times should you repeat each test? How will you measure and record the results? Do you need to find an average of your results?*

After the children have made and tested their parachutes, give them time to review their results.

Ask: *Are there any results that surprise you? Can you construct a line graph of your results? Which parachute took the longest to fall? What is the relationship between the variable you changed (the independent variable) and the time the parachute took to reach the ground? Can you explain this in terms of gravity and air resistance? Based on your results what improvement would you make to your first parachute?*

Challenge 3: Children evaluate how changing a measured independent variable affects the time it takes for their parachute to fall to the ground

Encourage the children to change a measured independent variable in at least three ways, for example, three different surface areas, in cm^2.

Ask: *Which variables will you keep the same (control)? How many times should you repeat each test? How will you measure and record the results? Do you need to find an average of your results? Are there any results that surprise you? Can you construct a line graph of your results?*

After the children have made and tested their parachutes, give them time to review their results.

Ask: *Which parachute took the longest to fall? What is the relationship between the variable you changed (the independent variable) and the time the parachute took to reach the ground? Can you explain this in terms of gravity and air resistance? Based on your results what improvement would you make to your first parachute?*

REFLECT AND REVIEW:

Ask the different groups to compare their findings so they can identify which parachutes took longest to reach the ground.

Ask: *What do you notice about the best parachutes? Can we be sure which are the slowest falling parachutes?*

Establish that it is impossible to say which parachute stayed in the air the longest unless all the parachutes were dropped from the same height. Then ask children to use their results to write a front page for *Parachute News* to explain what they have found out.

EVIDENCE OF LEARNING:

Watch and listen to children as they make and test their parachutes. Can children make a prediction based on their observations? Can they plan a comparative or fair test to investigate their prediction? Can they record their results clearly? Can they present their results in a bar or line graph, as appropriate. Do children identify that gravity pulls the parachutes to the ground? Can they explain that air resistance slowed the movement of their parachute? Were they able to link the surface area of the parachute to the time spent in the air? Do they report their findings using scientific vocabulary?

CROSS-CURRICULAR OPPORTUNITIES:

This activity makes a very good context for graph work and there are opportunities for line graphs, both of which link to Mathematics.

FEEL THE FORCE

LESSON 5: DOES THE SHAPE OF AN OBJECT AFFECT ITS MOVEMENT IN A LIQUID?

LESSON SUMMARY:

In this lesson children learn that water resistance is a form of friction that opposes movement in water. They explore how the shape of an object affects its movement through a liquid. By the end of the lesson they will are to describe how to reduce the effects of water resistance.

Key vocabulary:

water resistance, water, floating, ripples, drag, streamlined, surface area, float, sink, pull, force, gravity

Resources:

Modelling clay, viscous children's bubble bath, 1000 ml measuring cylinders (or 1.5 litre plastic bottles with the necks cut off), timers (stopwatches or stopclocks), 1000 ml jugs, digital scales, kitchen roll, sticky notes, elastic bands, masking tape (if required for the plastic bottles)

Health and safety:

If using plastic bottles with the necks cut off, stick masking tape over the cut edge of the plastic.

Key information:

Water resistance is the force that opposes any movement through or on the surface of water. It can be defined as the force of the water pushing back against an object moving through it. The more streamlined the object, the faster it moves and the less resistance it has.

National curriculum links:

Identify the effects of air resistance, water resistance and friction, which act between moving surfaces

Learning intention:

To measure the effects of water resistance

Scientific enquiry type:

Carrying out comparative and fair tests

Working scientifically links:

Taking measurements, using a range of scientific equipment, with increasing accuracy and precision, taking repeat readings when appropriate

Success criteria:

- I can take accurate readings using a stopwatch.
- I can use my results to identify which object has the least water resistance.
- I can identify the forces that start the object moving and stop the movement.

EXPLORE:

Ask children if it is easier to run in air or in water. Ask them to discuss their ideas in pairs, including reasons for their answers. Select some responses to share with the class.

Show children the Moving in water video (Video 1), which shows people on the beach. Ask children to look particularly at the boy who runs into the sea and the boy nearer the camera who is walking out of the sea. Prompt them to look at how the two boys move in the water.

Explain to children that it is a type of friction called water resistance that is making the children walk in that way in water. Water resistance is similar to air resistance that slows objects down as they move through air.

ENQUIRE:

Explain to children that they are going to investigate the best shape for the bow (front) of a speedboat. Ask them to discuss in pairs what they think might be the best shape and to draw the shape on mini whiteboards. Select some of their drawings and ask for their reasons for the choice of shape.

Explain to children that they are going to work in groups of three to test the shapes that they have drawn, and others, by dropping different modelling clay shapes into bubble bath liquid. The three challenges are differentiated by how much guidance and instruction each group is given, and by the collection and presentation of data required.

The challenges are presented on the Challenge slides to be displayed on the board, or printed out and placed in the centre of the table.

Challenge 1: Children measure the time taken for different bow shapes to fall though 700 ml of bubble bath liquid

Give each group of children a 1000 ml measuring cylinder (or a plastic bottle with the neck cut off), bubble bath liquid to fill the cylinder, modelling clay, two jugs, digital scales, kitchen roll, timers, sticky notes cut into thin strips and a selection of shapes on which to model their bows (Choose your shape, Resource sheet 1).

Ask the children to fill the cylinder to the 1000 ml (1 litre) mark with the bubble bath liquid and to place the top of a sticky note strip on the 900 ml mark on the cylinder. This is the timing start point. Then ask them to stick the top of another sticky note strip on the 200 ml mark. This is the timing stop point.

Then ask them to weigh four pieces of modelling clay of 25 g each, shape them into cuboids and make one end of each cuboid into one of the various shapes for the boat bows. Remind them to leave one cuboid with a flat end to use as the control shape. Encourage the children to write down their prediction of which shape they think will move through the bubble bath the fastest.

Then ask one child to hold the shape vertically, bow downwards, and partly submerged in the bubble bath mixture with the end of the bow held at the level of the start point. Ask a second child with a timer to tell the first child when to let go of the shape and to start the timer. As the bow of the shape passes the stop mark the second child stops the timer and the third child records the time in seconds in the Shape of your bow table (Resource sheet 2).

Ask the children to repeat the procedure for each shape. (There should be enough space at the bottom of the cylinder under the 200 ml mark for the three shapes without having to retrieve each shape after each drop.)

When they have completed their tests, ask the children to empty the bubble bath mixture into a jug, retrieve the four shapes, wash the shapes in a jug of water and dry them. They can then pour the bubble bath liquid back into the cylinder and top up to I litre if necessary.

Ask them to repeat the investigation, with the children swopping roles.

Challenge 2: Children measure the time taken for different bow shapes to fall though 700 ml of bubble bath liquid and devise a recording framework

Give the children the same resources as in Challenge 1 and ask them to set up their investigation in the same way as in Challenge 1. Ask them to select two shapes to investigate from the Bow shapes selection (Resource sheet 1), as well as their own shape and the control shape. Explain to the children that they should decide what to measure, how many times to make the measurements and how to collect their readings.

Challenge 3: Children devise an investigation to measure the time taken for different bow shapes to fall though 700 ml of bubble bath liquid

Give the children the same resources as in Challenges 1 and 2, including the Bow shapes selection (Resource sheet 1). Advise them that they should investigate at least three shapes as well as the control. Explain to each group of three children that they must decide how they are going to carry out the investigation so that they have comparable results, which they must record. They should decide on variables such as their start and stop points.

Ask: *How will you set up the investigation? What is the best way to record your observations? How will you decide the start and stop points? How many readings will you take? How can you explain any unexpected readings?*

Remind the children to record their results in an appropriate table.

Key information:

The more molecules that there are in a liquid, the thicker the liquid (i.e. the greater the effect of the resistance).

REFLECT AND REVIEW:

Ask each group to report its results and conclusion about which shape would be the best for a speedboat bow. Determine if all the groups found that the same shape fell fastest and if they think that the results on each shape from the different groups can be compared. If not, why not?

Ask children why they think bubble bath mixture was used instead of water. Demonstrate what happens when one of the modelling clay shapes is dropped in water. It drops very fast and would be almost impossible to time accurately.

Ask: *What does this tell you about resistance in water compared with resistance in bubble bath? Can you explain your ideas?*

Show children Boat bows video (Video 2) and prompt them to look at the shape of the bows.

Ask: *Is the speedboat's bow in the video the same as the fastest shape that you tested?*

EVIDENCE OF LEARNING:

Watch and listen to children as they carry out their investigations. Can they measure accurately using a timer? Can they record their results of repeated measurements in a table? Can they identify if any of their results are anomalous? Can they name the forces (gravity and resistance) that act on the modelling clay as it descends through the bubble bath mixture? Can children identify the most streamlined shape? Can they explain why this shape is streamlined, using the term 'water resistance'? Can children explain how water resistance can be reduced? Do they know that water resistance is the force that slows the movement of objects in water?

FEEL THE FORCE

LESSON 6: DO ALL HEAVY THINGS SINK?

LESSON SUMMARY:

In this lesson children measure the effect of upthrust on objects in water, by measuring and comparing weights of objects in water and air. They find out how the relationship between weight and size affects floating. By the end of this lesson they are able to use the term 'upthrust' to explain why something sinks or floats.

National curriculum links:

Identify scientific evidence that has been used to support or refute ideas or arguments

Learning intention:

To identify and explain the effect of upthrust on objects in water

Scientific enquiry type:

Carrying out comparative and fair tests

Working scientifically links:

Reporting and presenting findings from enquiries, including conclusions, causal relationships and explanations of and degree of trust in results, in oral and written forms such as displays and other presentations

Success criteria:

- I can explain why objects record a different mass in water and in air.
- I can record results systematically and take repeat readings.
- I can use the term 'upthrust' to explain how an object floats.

EXPLORE:

Start the lesson by showing the class the Floating or sinking slideshow (Slideshow 1).

Ask: *What forces do you think are acting on each object in the pictures?*

Remind children of previous learning about balanced and unbalanced forces.

Ask: *If the object isn't moving are the forces balanced or unbalanced? What forces are acting on these objects?*

Ask a child to make two identical weight balls of clay – about 100 g.

Ask: *What will happen when they are dropped into water?*

Ask if anyone thinks they could reshape the clay to make it float. If necessary, demonstrate how a hollow bowl shape will float.

Ask: *Why does the hollow ball shape float, but not the ball?*

Explain that the challenge is to find out why some objects float but others sink.

ENQUIRE:

Show children a 2.5 N Newton meter. Hang a ball of modelling clay from the Newton meter with an elastic band. Ask a child to read out its weight in Newtons. Lower the modelling clay ball into a tumbler half full of water until completely submerged. Ask the child to read out the weight in Newtons.

Write the two weights on the board. Let a different child repeat the measurements with a different sized ball of modelling clay. Ask children what they noticed about the modelling clay's weight each time. Then gently push upwards on the ball of modelling clay still attached to the Newton meter. Ask a child to watch the scale on the Newton meter and to say when it reaches the same reading as in the water. Invite a few children to take turns at pushing a small, inflated balloon into water in a tank.

Ask: *What can you feel?*

Children should feel the push of the water. Explain that this force is called upthrust.

The challenges are differentiated by type and precision of recording and complexity of context. Evidence from all three challenges will be combined to explain why things float. Children work in groups of three.

The challenges are presented on the Challenge slides to be displayed on the board, or printed out and placed in the centre of the table.

Challenge 1: Children investigate what happens when fruit and vegetables are put in water

Give each group a selection of whole fruit and vegetables. Ask them to predict and record which will float and which will sink. Ask them to put each fruit or vegetable in the water and compare what happens with their prediction.

Ask the children to record what happens when three whole fruits and/or vegetables, and then different sized pieces of the same fruits and vegetables, are put in the water. Ask them to draw a diagram of a floating fruit and a sinking fruit using arrows to show the forces at work. Remind them that force arrows show direction and size.

Challenge 2: Children investigate the weight of objects in air and in water

Give each group balls of modelling clay of different sizes. Ask them to form the clay into different shapes, including some bowl shapes, and to place an elastic band around each shape so they can hook it onto the Newton meter.

Ask the children to record the weight in Newtons of each ball of clay, first in air and then in the tumbler of water. Ask them to mark on the outside of the tumbler how far the water rises for each ball of clay.

Ask the children to record their results in Effects of upthrust table (Resource sheet 1) and to draw a diagram to show what they found out, using force arrows for gravity and water resistance. Remind them that force arrows show direction and size. Ask if anyone made a clay shape that didn't appear to weigh anything in water. If so, ask why they think this happened.

Challenge 3: Children investigate the effect on upthrust of adding salt to water

Ask children to record the weight in Newtons of modelling clay in air, fresh water and salt water (about two teaspoons of salt per tumbler of water). Provide large plastic tumblers of fresh water and salt water. Ask them to use different amounts of clay, but always in a ball shape.

Ask the children to decide how to record the results for air, fresh water and salt water and, before submerging the modelling clay in salt water, to record their prediction for the reading on the Newton meter. Ask the children to explain what they think is happening from the readings they have taken and to draw a diagram to show what they found out, using force arrows for gravity and water resistance. Remind them that force arrows show direction and size.

REFLECT AND REVIEW:

Ask children what happened when they put different sized pieces into water.

Ask: *Did the pieces do the same as the whole fruit or vegetable – sink or float? Did the bigger things do the same as the smaller things? What about heavier things? What did you notice?*

Establish that things that were heavy for their size sank and things that were light for their size floated.

Ask children what happened to the water level as each piece was submerged.

Ask: *Did heavier or larger shapes make the water rise most? What did you notice about the weight and size of the objects that floated?*

Establish that upthrust from the water pushed up on the objects, reducing the weight (the downward pull of gravity). Ensure that children understand that larger objects caused the water to rise more: they moved (displaced) more water.

Ask children how adding salt to water affected the weight of an object suspended in it. Establish that the upthrust on the clay balls was greater in the salt water.

Show the different clay shapes from the beginning of the lesson.

Ask: *How do your investigations explain why the ball of clay sank and the bowl shape floated?*

You may need to help children to develop an explanation (see Key information).

EVIDENCE OF LEARNING:

Watch children as they work, look at their diagrams and listen to their explanations of what they have observed. Do they use appropriate methods of recording their findings? Do they provide clear descriptions about what happens when things are placed in water? Do they use the term 'upthrust' correctly in their observations? Do they show on their drawings that gravity is the force that pulls on objects suspended on the Newton meter? Can children explain that in water, some of the object's mass is supported by a force called upthrust? Do they use evidence from their investigations to explain why objects sink or float?

FEEL THE FORCE

LESSON 7: HOW FAR CAN YOU STRETCH?

LESSON SUMMARY:

In this lesson children investigate what happens to rubber bands and springs when a force is applied. By the end of the lesson they are able to use their results to describe the relationship between the force applied and the distance the spring or band stretches. The lesson brings together learning from across the module so far, including understanding about the downward force exerted by gravity on objects (weight) and the measurement of forces.

Preparation required: You need to prepare a fixing point on which to hang a spring, which has sufficient drop to hang slotted masses of up to 3kg (for Challenge 3).

Key vocabulary:

pull, stretch, extend, push, Newton meter, weight, gravity

Resources:

Fixed cup hooks, selection of springs, paper clips, rubber bands, sets of hanging slotted weights up to 100 g, hanging weights of 50 g and 100 g, Newton meters, tape measures, modelling clay, small objects from around the classroom to hang on rubber bands

Health and safety:

Be aware of falling objects. It may be necessary to wear eye protection when stretching rubber bands. See Be Safe! section 13.

Key information:

Mass is the measure of the amount of material in an object. Weight is the gravitational force acting on a mass. This is measured in Newtons.

National curriculum links:

Explain that unsupported objects fall towards the Earth because of the force of gravity acting between the Earth and the falling object

Learning intention:

To explain why a larger mass stretches a rubber band or spring more than a smaller mass

Scientific enquiry type:

Noticing patterns

Working scientifically links:

Recording data and results of increasing complexity using scientific diagrams and labels, classification keys, tables, scatter graphs, and bar and line graphs

Success criteria:

- I can carry out a test to investigate how mass affects the amount that springs or rubber bands stretch.
- I can identify a pattern in my results.
- I can explain how springs help us to weigh objects.

EXPLORE:

Give four children rubber bands of different lengths and thicknesses. Use hooks or paper clips to hang small objects from around the classroom on the bands.

Ask: *What has happened to the rubber bands? What is pulling them? Have they all stretched the same amount? Why do you think this is?*

Remind children that it is not the objects themselves that are pulling the bands down, but the force of gravity acting on the objects. Explain that weight is how we measure the force of gravity pulling on an object.

Next, give four different children springs of different sizes. Hang the same objects on the ends of the springs.

Ask: *What has happened to the springs? Have they all stretched the same amount? Why do you think this is? How have they behaved differently or similarly to the rubber bands?*

Establish that the amount a rubber band or spring stretches is affected by its thickness, length and the material it is made from, as well as the amount of the mass that is hanging on it.

ENQUIRE:

Explain to children that they are going to investigate how far rubber bands and springs stretch when different masses are added, and that they will use that information to compare the strength of the bands and springs and the weights of different objects. The challenges are differentiated by how much support children are given to carry out the test and the type of data presentation and interpretation that is required.

Group children into pairs according to their contributions to the discussion in the Explore section and particularly their understanding about gravity exerting a downward force that stretches the rubber band or spring.

The challenges are presented on the Challenge slides to be displayed on the board, or printed out and placed in the centre of the table.

Challenge 1: Children measure how far different springs stretch and rank them

Give each pair of children six different springs to test. Explain to them that their task is to compare how far the different springs stretch, using the same mass on each spring that they test. Ask each pair to suggest how they can do this. Encourage them to create their own table to record how far the springs stretch. Ask them to put the springs in order of how far they stretched (extension) and to write a sentence to summarise what they found out.

Challenge 2: Children investigate how far a rubber band is stretched by different masses

Ask the children, in pairs, to investigate how far a rubber band stretches when regularly increasing masses are hung on it. Ask the children to write a question for their investigation (Resource sheet 1). Explain that they should use a table to record their results and ask them to draw a line graph to present the data showing how far the band stretched. Give them the Rubber band stretch table (Resource sheet 2) to help them. Explain that they should also write a sentence to explain what their data shows.

Challenge 3: Children investigate the extension of a single spring and independently construct a line graph to present their results

Ask the children, in pairs, to investigate how much a spring stretches when different masses are hung on it. Explain that they should hang regularly increasing masses on the spring and record in a table the extension of the spring with each mass. Ask them to transfer this information to a line graph, with the mass along the bottom of the graph (horizontal axis) and the extension length up the side (vertical axis). Ask them to write a sentence that describes the relationship between the mass and the distance the spring stretches.

REFLECT AND REVIEW:

Ask three pairs of children who have completed the different challenges to read out the sentences that summarise what they have found out. Does everyone else agree with their sentences?

Display one of the Challenge 3 graphs, which should show the direct relationship between the mass hung on the spring and the amount the spring stretches. Give a child a ball of modelling clay and ask them to estimate its mass. Then ask them to find its actual mass by measuring how far the spring extends and using the graph to read the mass required to extend the spring by the measured distance.

Ask: *Does this remind you of anything?*

Give children a few minutes to look closely at the Newton meters that they have used in previous lessons.

Ask: *What is the working part made from? Can you use the graph you have just interpreted to explain how it works?*

EVIDENCE OF LEARNING:

Listen to children as they carry out their investigations, look at their graphs and written sentences, and listen to what they say about how the Newton meter works. Do they plan how to carry out their investigation systematically? Can they measure accurately? Do they take repeat readings? Are children able to describe the relationship between the mass and the extension of the spring or band? Can they use the line graph to find the mass of the modelling clay? Can they explain how a Newton meter works?

FEEL THE FORCE

 ## LESSON 8: HOW CAN WE USE LEVERS TO HELP US?

LESSON SUMMARY:

This lesson introduces mechanisms – devices that change the effect of a force. Children investigate levers for moving things and increasing/decreasing a force. By the end of the lesson children know that levers can alter the size of forces required to move or balance an object.

Key vocabulary:

lever, pivot, push, pull, mechanism, machine, force, fulcrum

Resources:

Everyday objects that use class one levers (for example, claw hammer, scissors, pliers, metal spoon), empty tins with inset metal lids, long-handled wooden spoons, 1 litre plastic bottles filled with water to weigh 1 kg, stiff cardboard tubes approximately 3 cm diameter, modelling clay, push/pull meters up to 10 N, books (at least thick paperback size), wooden ramps to be used in Challenge 1

Key information:

A mechanism is simply a device that takes an input motion and force, and outputs a different motion and force. A lever is the simplest kind of mechanism.

National curriculum links:

Recognise that some mechanisms, including levers, pulleys and gears, allow a smaller force to have a greater effect

Learning intention:

To demonstrate how levers work and how they reduce the force required to move objects

Scientific enquiry type:

Carrying out simple comparative and fair tests

Working scientifically links:

Taking measurements, using a range of scientific equipment with increasing accuracy and precision, including taking repeat readings when appropriate

Success criteria:

- I can use levers to move objects.
- I can explain that when something is moved using a lever, less force is needed.
- I can label a diagram using scientific vocabulary to explain how a lever works.

EXPLORE:

Ask children what they think a mechanism is. Explain that it is a device that makes work using forces easier. Inform children they are going to use levers – a type of mechanism.

Show everyday levers: scissors, pliers and claw hammer. Explain that these levers need a moving force to make them work. Give children empty tins with inset lids. Ask them to work in pairs to open the tins. Ask if they know how to open them easily.

Give out the teaspoons and ask children to use them to open the tins.

Ask: *How did you use the spoon to open the tin?*

Emphasise that they used a force to push down on the spoon, which pivots on the edge of the tin. There is an upward force on the other end of the spoon that lifts the lid. (Put the end of the teaspoon handle, not the bowl of the spoon, under the lip of the lid.) Remind children of the arrows they used previously to show directions of forces.

Ask: *In which directions are forces working? Where is the pivot point? What playground item works like this?*

To explain the lever mechanism ask a child to hold the end of the handle of a long-handled wooden spoon. Explain that they are representing the lid of the tin. Hold the handle between your finger and thumb near the same end of the spoon. This is the fulcrum, where the lever pivots. Show what happens when a downward force is applied to the other end of the spoon as it pivots in between your finger and thumb. Ask the child what they feel happening where they are holding the handle.

Establish that the spoon is a lever and that this was the mechanism that helped to open the tin.

Explain that levers have three parts: the fulcrum, where the lever pivots; the weight arm (the part from the fulcrum to the weight to lift); and the force arm (the part from the fulcrum to where you push or pull).

Ask: *Can you name these parts in the lever you used to open the tin?*

Ask children each to draw on mini whiteboards a diagram of the tin being opened with a spoon and label it with this vocabulary.

ENQUIRE:

Explain to children that they are going to investigate how to use levers to change the force needed to move things. The challenges are differentiated by how children describe or measure the effects of levers. Children work in groups of three.

The challenges are presented on the Challenge slides to be displayed on the board, or printed out and placed in the centre of the table.

Challenge 1: Children investigate the effects of the comparative lengths of lever and weight arm on moving a load

Give each group two tins with inset lids (to act as the fulcrum), a wooden board, books and Resource sheet 1. Tell them to tape the tins together end to end, rest the board across them like a seesaw and place four books on one end. This is the weight arm.

Ask them to add books to the pile, one at a time until the weight arm lifts, and record the number of books. Ask them to repeat this twice: first making the force arm longer than the weight arm, then making both arms the same length. Encourage the children to write a sentence saying which makes it easier to move the weight: a force arm is shorter, longer than, or same length as, the weight arm.

Challenge 2: Children measure the difference a lever makes to the force needed to move an object

Explain to the children that they going to investigate the effect of changing the length of a lever. Give each group a long-handled wooden spoon, a 1 litre plastic bottle filled with water (i.e. 1 kg), a cardboard tube, modelling clay and a push/pull meter.

Provide Get things moving (Resource sheet 2). Ask the children to place the lever (spoon) under the neck of the bottle lying on its side, just below the lid, and measure and record the force (in Newtons) needed to lift the bottle with the fulcrum (cardboard tube) in four different positions: close to the bottle, and then moving it backwards towards the bowl of the spoon. Suggest that one child holds the modelling clay to stop the bottle rolling, another pushes down with the push meter and the third takes the reading from the push meter. Ask them to record these on mini whiteboards.

Challenge 3: Children measure the difference a lever makes to the force needed to move an object

Give each group a long-handled wooden spoon, a 1 litre plastic bottle filled with water (i.e. 1 kg), a cardboard tube, modelling clay and a push/pull meter. Explain to the children that they going to investigate the effect of the length of a lever.

Provide Get things moving (Resource sheet 2). Ask the children to measure and record the force (in Newtons) needed to lift the bottle, and record their measurements on the Bottle lift table (Resource sheet 3). Ask them to place the lever (spoon) under the neck of the bottle (lying on its side) just below the lid, with the fulcrum as close to the bottle neck as possible, and to push down on the bowl of the spoon with a push meter. Suggest that one child holds the modelling clay to stop the bottle rolling, another pushes down with the push meter and the third takes the reading from the push meter. Ask them to repeat this but to move the fulcrum away from the bottle neck at intervals and take and record readings until they reach the bowl of the spoon.

REFLECT AND REVIEW:

Prompt children to look again at their tin opening diagrams.

Ask: *Would it be easier to open the tin with the fulcrum further away from the lid?*

Ask Challenge 1 children how many books were needed to lift the weight at each position of the fulcrum. Ask the class if their results on lifting the bottles show that the force needed to lift the weight decreases as the fulcrum moves towards the weight.

EVIDENCE OF LEARNING:

Observe children during challenges and look at their recording. Do they use the terms 'fulcrum', 'weight arm' and 'force arm'? Can they explain how levers work? Do they understand that the force needed to move the weight decreases the closer the fulcrum is to the weight? Do they record results accurately? Do they use results to support explanations?

FEEL THE FORCE

LESSON 9: HOW CAN WE LIFT A HEAVY LOAD?

LESSON SUMMARY:

In this lesson children use pulleys to lift objects. By the end of the lesson they understand that pulleys make lifting easier by reducing the force needed.

Preparation required: Set up hanging points for lengths of dowel, or brush handles, from which to hang pulley sets or coat hangers. A pair of chairs would suffice if they have holes in the tops of the backrests. Weight the chairs with books on the seats or rest the dowel or brush handles between two tables and secure them with masking tape and weight them with books. Test that the string used with pulleys is strong enough to lift the objects.

National curriculum links:

Recognise that some mechanisms, including levers, pulleys and gears, allow a smaller force to have a greater effect

Learning intention:

To explain why pulleys make lifting objects easier

Scientific enquiry type:

Noticing patterns

Working scientifically links:

Reporting and presenting findings from enquiries, including conclusions, causal relationships and explanations of and degree of trust in results, in oral and written forms such a displays and other presentations

Success criteria:

- I can explain how a pulley works.
- I can plan and carry out an investigation to find out how a pulley affects the force needed to lift a load.
- I can notice a pattern between the number of pulleys used and the force needed to lift a load.

Key vocabulary:

pull, lift, force, effort, mechanism, machine, pulley

Resources:

Wooden dowel (at least 2 cm diameter) or brush handles, pulley sets or metal coat hangers and curtain rings to slide on dowel, cotton reels, string or thin rope, small bucket, sand, Newton meters, sets of slotted weights in 10 g denominations, hanging masses of 50 g, 200 g, 500 g and 1000 g

Health and safety:

Place a box of soft loose fabric or crumpled paper under objects when lifted. Ensure that pulleys are attached to a secure fixing. See Be Safe! section 13.

Key information:

Both lifts are examples of simple pulleys, but the lift over the dowel might give a higher force reading than the straight lift because of friction between string and dowel.

[handwritten note: Brush handle, String cotton reel newton meter]

... how levers help with lifting objects. Ask them

... te another type of mechanism that helps with ... te a simple pulley: lift a small bucket of sand ... ure the force with a Newton meter. Lift it again ... ach end. Measure the force. Finally, thread the ... h the string over the cotton reel. Measure the

... call this mechanism. Prompt them to think of anywhere they have seen pulleys used. Ask what they were used for.

If the classroom has blinds demonstrate how pulleys are used to lift and lower them.

Show children the Cranes video (Video 1).

ENQUIRE:

Demonstrate lifting a 500 g hanging weight with two pulleys (Resource sheets 1, 3 and 4 give a pictorial setup). Ask a child to read the force required. Add 20 g of slotted weights to the 500 g and repeat the lift, with a child measuring the force. Ask if increasing the weight by 20 g at a time gives a meaningful reading, and if not, ask what the increase in weight should be. (You might need to suggest starting with 500 g and going up in 250 g increments.)

Explain to children that they are going to measure how pulleys make lifting loads easier. Ask them to measure accurately in Newtons.

Ask children why they should measure the force times for each weight. If necessary, remind them that taking repeat readings increases their reliability as any can ignore any that are very different from the majority.

Key information:

Challenges can be completed using curtain rings and coat hangers as in the Pulleys setup sheet (Resource sheets 1, 3 and 4), but commercially produced pulley sets are recommended. Screw cup hooks into the dowel or brush handle from which to hang pulley sets if they won't hook over large diameter dowel. For curtain rings and coat hangers hang the load from a coat hanger for comparison of results with the other two groups. Slightly bend the hanger outwards to stop the load string sliding.

The challenges are differentiated by the complexity of the mechanism, for example, the number of pulleys tested, but each group uses the same types of pulleys and string. Children work in groups of three.

The challenges are presented on the Challenge slides to be displayed on the board, or printed out and placed in the centre of the table.

test 80g
10–100g w/ +
w/ Pulley record

Challen... two pulleys

Give th... them to set up two pulleys and predict... nvestigate 100 g, 250 g, 500 g, 750 g a... t weights from the floor to the top pulley, a... their results in the Pulley system results t... w to make their results as reliable as possi...

Challen... three pulleys

Give the... them to set up three pulleys and predict t... vestigate 100 g, 250 g, 500 g, 750 g a... force needed to lift weights from the floor to the top pulley, and the distance they pull the string. Ask them to record results in the Pulley system results table (Resource sheet 2). Remind children to consider how to make their results as reliable as possible.

Challenge 3: Children investigate lifting different weights with four pulleys

Give them the Four pulley system sheet (Resource sheet 4). Ask them to set up four pulleys and to predict the force needed to lift different weights. Ask them to investigate 100 g, 250 g, 500 g, 750 g and 1000 g. Ask them to measure the force needed to lift weights from the floor to the top pulley, and the distance they pull the string. Ask them to record results in the Pulley system results table (Resource sheet 2). Remind children to consider how to make their results as reliable as possible.

REFLECT AND REVIEW:

Key information:

The results should produce a straight line graph. The line might curve as friction increases.

Prepare a results table on the board for each group to complete with results from their two-, three... sheet (Resource sheet 5) as a template.

Ask... ed for the same weights using the diffe...

Ask:...

Expl... graph. Use the template in Resource shee...

Disc... tion on the line graph to write a sent... to lift different weights.

Have axis ready on board. Plot as class

EVI...

Listen to children during the challenges and as they ... their results. Do they make predictions about forces needed to lift weights? Do they measure the force and the distance the string has travelled accurately? Do they repeat readings? Do they understand that as the number of pulleys increases so does the distance they pull the string, but the force used decreases? Can they interpret data to identify a pattern between the force required to lift a weight and the number of pulleys?

FEEL THE FORCE

LESSON 10: CAN A WHEEL WITH TEETH MAKE WORK EASIER?

LESSON SUMMARY:

In this lesson children learn about gears, a third type of mechanism that helps us to do things by changing the effect of forces. Children identify where gears are used in everyday life. By the end of the lesson they understand how gears are used to transfer movement from one place to another and how they can change the speed and direction of movement, and convert a small force into a bigger action.

Key vocabulary:

gears, forces, cogs, wheels, teeth

Resources:

Balloon whisk, rotary whisk, 2 bowls and 2 egg whites, cheap clock with removable back, commercially produced plastic gear wheels, plastic bricks, axles, plastic brick bases

Health and safety:

Children should work with the equipment under supervision. Remind children to be careful with the kitchen items. Ensure children do not use sharp items without supervision.

Key information:

Gears are an example of a mechanism. A gear is a wheel with teeth. The teeth are also called cogs and gears are sometimes called cog wheels. One gear on its own is not any use; you need to have at least two gears working together. Because the teeth fit together, when one wheel is turned the other one also turns, but in the opposite direction. Gears can increase the amount of work accomplished, for example, if you have a big gear attached to a small gear, when the big wheel is turned slowly the small gear turns quickly. This means gears can be used to reduce the amount of effort needed. Most children will have seen or heard about gears in their bikes.

National curriculum links:

Recognise that some mechanisms, including levers, pulleys and gears, allow a smaller force to have a greater effect

Learning intention:

To explain how gears allow a smaller force to have a greater effect

Scientific enquiry type:

Noticing patterns

Working scientifically links:

Recording data and results of increasing complexity using scientific diagrams and labels, classification keys, tables, scatter graphs and bar and line graphs

Success criteria:

- I can identify gears in a range of everyday items.
- I can make simple gears and explain how they work.
- I can explain how gears convert a force.

EXPLORE:

Ask children if they [...]

Ask: *What are they [...] and Technology? Do you have a bi[...]*

Explain that a gear [...] ary egg whisk and a balloon hand whisk [...]

Ask them how man[...] to think about which way the gears mov[...] turned.

Ask two children to [...]isk and one using a rotary whisk. Expla[...] and then swap whisks. Ask them which w[...] e machine (the rotary whisk) made it eas[...]

[handwritten note: Compare Measured Whisk to rotary whisk*]*

Explain to children that gears make some things easier to do by changing the force needed to make something move. Remind them about the other mechanisms that they have learned about in the module that help us to do things.

Ask: *How are gears different from levers and pulleys?*

Show the video of the cyclist (Video 1). Ask children to watch how the speed of the pedalling changes when he is cycling along a flat road, up a hill and in a time trial. Explain to them that increasing the size of one gear, where the pedals are attached, makes the smaller gears on the back wheel work faster.

ENQUIRE:

Explain to children that they are going to investigate how gears work in different combinations. Introduce them to the idea that the number of teeth and the size of the gear affect how many times the gear rotates when it is meshed with another gear of a different size. Inform children that they need to make careful observations to count the number of teeth on the different gears and the number of turns, and the changes to the direction of movement that occur when the gears are combined. The challenges are differentiated by the way children are required to record the changes to the movement that they observe and the complexity of the mechanisms that they construct and describe. Children work in groups of two or three.

The challenges are presented on the Challenge slides to be displayed on the board, or printed out and placed in the centre of the table.

Challenge 1: Children investigate how gears move

Give each group commercially produced plastic gears of different sizes. Ask them to see if they can construct a mechanism that uses at least three different sized gears with different numbers of cogs. Challenge the children to count the number of turns of the individual gears and the direction of movement. Ask them to think about how the number and the size of the gear affect how many [turns] of a different size.

Ask the [children to draw labelled diagram]s and labels, and ask them to write an 'If... [then... because...' sentence about] its rotation.

Challenge [2: Children investigate how gears speed up and slow] down motion

Give ea[ch group ... plastic gears of differe]nt sizes. Explain that their challenge is to us[e ...] [the relationship betw]een the size of gear and the number of turn[s ...] [diagram]s) that show that when one gear (the dri[ving gear] ...) [one gear] (the driven gear) in the sequence turns th[e ...]

Ask the [children to draw labelled diagrams sh]owing the number of rotations of each ge[ar ... direction in which each] gear turns. Ask them to write two 'If...then...because...' sentences about what they have done.

Challenge 3: Children use different types of gears to change the speed and direction of the motion

Give each group commercially produced plastic gears of different sizes. Explain that their challenge is to use at least four gears to generate a gear train for which the speed of the final cog should be half the speed of the first cog, and the final cog should turn in a different plane to that of the first cog.

Ask the children to record the gear ratios by counting the number of cogs on the gears and to draw labelled diagrams showing the number and direction of turns. Ask them to write two 'If...then... because...' sentences about what they have done.

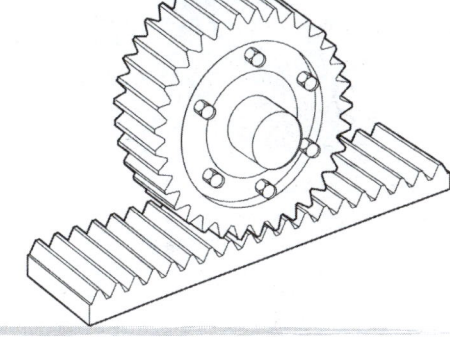

Handwritten note (yellow sticky):
1. Cogs + cotton reel
2. Building toy cogs (LMs)

Key information:

Gear ratios are dependent on the number of cogs (teeth) on a gear. If a driving gear has 10 teeth and the driven gear has 20 teeth then the driving gear turns twice for one turn of the driven gear – a gear ratio of 2:1. Adding larger gears slows down the motion and smaller gears speed it up.

Key information:

If the driving gear is set up vertically then the final driven gear must be horizontal and vice versa.

REFLECT AND REVIEW:

At the end of the Challenge section give children a few minutes to have a look at other children's mechanisms. Ask them to identify three ways that gears affect motion and to write them in their books. Bring the children back together to share their findings.

Establish that gears can make things move faster; can change the direction of movement; and can reduce the amount of force that is needed.

Give children a variety of everyday items that use gears, for example, a tin opener, a clock and the rotary egg whisk.

Ask: *How many gears can you see? In what way do these gears change the movement of the object?*

Show children the Clock video (Video 2) and ask them to watch how the gears move. Extend this into a homework activity for children to identify an item in their home that has gears and decide how they work.

EVIDENCE OF LEARNING:

Watch and listen to children in their groups as they carry out the challenges. Review the diagrams and the 'If...then...because...' sentences as evidence of children's understanding of how gears change speed and direction. Do children understand that the size of gears affects the speed of other gears and how gears change the direction of movement? Do they understand that a large gear makes a smaller one rotate much faster, and vice versa? Can they identify gears that are used in everyday life?

CROSS-CURRICULAR OPPORTUNITIES:

This lesson links to D&T as children can use their understanding of cogs and gears to create a model that moves.

THE EARTH AND BEYOND

INTRODUCTION

In this module children develop their knowledge of the Earth's (and other planets') place in the solar system, and their relationships with other bodies in space, in particular with the Sun.

The module draws on children's previous use of the calendar to calculate the duration of events (Year 4 Mathematics) and solve problems involving units of time (Year 5 Mathematics). Children also learn how the Earth's orbit determines the length of a year and why we have leap years.

Key Stage 1 observations of the Sun's movement across the sky and Year 3 work on shadows provide a sound basis for investigating how the Earth's rotation causes night and day, and is responsible for the apparent movement of the Sun across the sky, and its changing height in the sky. Children also learn how the Earth's rotation and tilt affect the direction and length of shadows, and how to use shadows for telling the time.

Children learn about time differences around the world and investigate time differences using resources including the internet. They will find out about how time was standardised around the world, about the need for scientists to choose a starting point in the continuous process of cycles of sunrise and sunset, and investigate longitude. They are introduced to the International Date Line and the Greenwich Meridian.

Children extend their awareness of seasonal changes through the year, which they developed during Key Stage 1, to understand that it is the Earth's tilt on its axis that causes the seasons. This draws on their learning about the Sun and shadows to develop an understanding of the role of latitude in day length and seasons.

When working scientifically, children use models for exploring and demonstrating ideas, first-hand observation made at night either in their gardens or local area, or from visits to local observatories, secondary sources of information (mainly web-based) to answer scientific questions increasingly independently, and diagrams, charts and graphs for recording data. They report and present findings in different ways, including booklets, oral presentations and annotated diagrams, draw conclusions, identify causal relationships and explain their thinking.

Before beginning work on this module, give children the Moon diaries instructions (Resource sheet 1 for Lesson 8). They look at the Moon each night, facing south, draw its shape and check, if the Moon is not a complete circle, which side is rounded and which side has a piece 'missing'.

Before Lesson 4 children should make shadow clocks, as described in the instruction sheets for that lesson (Resource sheets 1 and 2), and record the positions and lengths of the shadows.

Key vocabulary:

Aldebaran, Arctic, Antarctic, British Summer Time, Earth, Greenwich Meridian, International Date Line, Jupiter, Mars, Mercury, Milky Way, Moon, North Pole, Saturn, South Pole, Sun, Neptune, Universe, Uranus, Venus, asteroid, autumn, axis, compass, crescent, dawn, degrees, dusk, equator, equinox, fixed stars, Full Moon, galaxy, gibbous, hemisphere, horizon, illuminate, leap year, longitude, lunar month, meridian, nebula, New Moon, northern, orbit, planet, reflect, rotate, rotation, solar system, solstice, southern, spin, spring, star, summer, sunrise, sunset, telescope, temperature, tilt, time zone, waning, waxing, winter, year, change, compare, draw conclusions, explain, explanation, investigation, line graph, measure, model, observations, plan, predict, prediction, presentation, question, record, review, scientific diagram, table

FACT FILE:

Stars are held together in a galaxy by gravity. (Our star, the Sun, is in the Milky Way galaxy.) 'Constellation' is not a scientific term but is commonly used for a pattern of stars in a clearly defined area of the sky. These stars may be vast distances apart and in different galaxies. The stars are called fixed because they were long believed not to move. Galaxies rotate: the distances between them are so great that this can only be detected using modern scientific equipment. The sky today looks as it did thousands of years ago and many of the constellations were named in ancient times.

When viewed from above the North Poles of the Sun and the Earth, the Earth and other planets orbit the Sun anticlockwise, causing an apparent shift in the positions of the stars over the year. This is not to be confused with the apparent rotation of the stars around the North Star Polaris, which is caused by the Earth's rotation.

The ancient Sumerians (3rd millennium BCE) based time around the number 60 – the smallest number divisible by every number from 1 to 6. This simplifies fractions based on 60: an hour can easily be divided into segments of 30, 20, 15, 12, 10, 6, 5, 4, 3, 2 and 1 minute. The measurement of angles is based around 60 for the same reasons, making the analogue clock face ideal for measuring time.

The cereal box clock children make in Lesson 4 is based on an ancient Egyptian sun clock from before 1500 BCE, which was rotated once a day at noon in order to tell the time in both morning and afternoon.

Places that are close to each other have sunrise and sunset at different times so, historically, cities and villages agreed local times, but this caused problems as industrialisation and rail travel developed. Great Britain and Ireland adopted the local time of London (Greenwich) as standard time. In the United States and Canada the times of sunrise and sunset differ enormously across the country, so railway companies specified junctions where changes in time were made.

Increased communication and travel brought the need for global agreement on dates and times. An international congress met in Washington DC in 1884 where the world was divided into 24 time zones (one for each hour of the day), each covering 15° of longitude. The time for each zone is that of the meridian (line of longitude) that passes through its centre. As the Earth's rotation, and with it sunrise and sunset, is a continuous process, a starting point was established. This is the meridian that passes through the Greenwich Observatory (the zero or prime meridian).

Children might think seasons occur because the Earth is nearer to the Sun in the summer and farther away in winter. The slight variation in the Earth's distance from the Sun, due to the Earth's elliptical (rather than circular) orbit, is not the cause. It makes no detectable difference because of its vast distance from the Sun. The tilt of the Earth's axis angles either the northern or southern hemisphere towards the Sun in the summer and away from it in winter, with midway points in spring and autumn. Day and night are the same lengths on the equinoxes: September 22nd/23rd and March 20th. The solstices are the 'longest' and 'shortest' days (with the longest or shortest period of daylight): June 20th/21st and December 22nd/23rd. Arctic and Antarctic regions have 24-hour, or almost 24-hour, daylight or night. Days and nights in tropical and equatorial regions are equal, or almost equal, in length throughout the year. Daylight Saving, in which clocks are adjusted by an hour (for example, British Summer Time), affects times in many countries.

Common misconceptions:

- Children often think that day and night are caused by the Earth orbiting the Sun (or even the Sun orbiting the Earth, rather than the Earth's rotation on its axis).

- Children may think that stars are 'star-shaped' with five points.

- Children may think that the Moon gives out light – actually it reflects light from the Sun.

- Children may consider the Universe to be the same thing as the solar system, with no other suns or planets; a solar system is a star with planets orbiting round it; a galaxy consists of hundreds of billions of stars, all of which are potential solar systems.

- Children may think that the change in shape of the Moon during the course of a month is to do with light being blocked because of shadows, when it is actually caused because the portion of the Moon and the Sun's reflection on it keeps changing, so we see it part-illuminated.

Big Cat book links

Black Holes Anna Claybourne 978-0-00-746540-8 Band 14 Ruby	What is a black hole? Where do they come from? What do they do?
The Big Bang Andrew Solway 978-0-00-733641-8 Band 16 Sapphire	What was the Big Bang? What was the Universe like just after the Big Bang? Who discovered it?
The Traveller's Guide to the Solar System Giles Sparrow 978-0-00-723101-0 Band 16 Sapphire	Bored spending all your holidays on Mars? Travel further into space with this fact-packed guide.
Early Space Scientists: Copernicus, Galileo and Newton Katie Daynes 978-0-00-753017-5 Band 18 Pearl	Today, space scientists have built amazing new telescopes and satellites to look at galaxies millions of light years away from Earth. But it's the work and ideas of early space scientists like Copernicus, Galileo and Newton that have influenced how we see space, the universe and beyond ...

THE EARTH AND BEYOND

LESSON 1: WHAT'S IN SPACE?

LESSON SUMMARY:

In this lesson children make observations of the night sky. Using secondary sources of information they consider explanations for, and raise questions about, their observations. They find answers to some of their questions through a 'journey into space', during which they explore diagrams and photographs of the solar system and beyond.

By the end of this lesson they know about the shapes and positions of the Earth and other planets in the solar system, their relative sizes and orbits around the Sun. They have some idea about where our solar system is in the Universe.

Preparation required: Cut out enough true or false statement cards (Resource sheet 1) for one set per group of three. Prepare star question shapes (Resource sheet 2) so that pupils can write their questions for display on a question board.

Key vocabulary:

asteroid, crescent, Earth, galaxy, Jupiter, Mars, Mercury, Milky Way, Moon, orbit, planet, Saturn, solar system, star, Sun, sunrise, sunset, Neptune, telescope, Uranus, Venus

Resources:

A2 paper

National curriculum links:

Describe the movement of the Earth and other planets in the solar system relative to the Sun

Learning intention:

To describe the shapes, positions and movement of the planets in the solar system and some of the differences between these and stars

Scientific enquiry type:

Finding things out using a wide range of secondary sources of information

Working scientifically links:

Recording data and results of increasing complexity using scientific diagrams and labels, classification keys, tables, scatter graphs, and bar and line graphs

Success criteria:

- I can ask questions that help me to find out about the solar system.
- I can use secondary sources of information to answer my questions.
- I can recognise that the Earth and other planets and the Moon are spheres.
- I can describe how the Earth and other planets move around the Sun.
- I can identify the Sun as a star.

EXPLORE:

Split the class into groups of three. Give each group a set of Space cards (Resource sheet 1) and some poster putty. Mark a sheet of A2 paper with headings True, False and Not Sure. Ask children to look at the cards, discuss the statements and decide which are true, which are false and which they are not sure about. Children fix their cards in their sets using poster putty. This provides an indication of their knowledge and ideas.

ENQUIRE:

Explain that they are going to look at pictures of the sky and make an imaginary journey into space. Display slide 1 of Journey into space (Slideshow 1) – a photograph of the night sky.

Ask: *What can you see? Is it night or day? How can you tell? What makes it dark? What will make it light there?*

Prompt children to think about the shapes of the objects in the sky, their colours, brightness and sizes. Children should notice the darkness, the Moon and stars, but might not realise that some of the 'stars' are planets.

Show slide 2 of Journey into space – the same photograph with some night sky objects labelled. Point out the visible planets and stars. Invite children to contribute to a list of objects in the picture in order of size, from the smallest to the largest, including trees and buildings.

Ask: *Are these bigger or smaller than stars? Are they bigger or smaller than planets? If they look bigger, why?* (They are nearer.) *What makes the stars and planets look smaller?*

Encourage children to consider how far away the objects in the night sky are.

Ask: *Do you think that planets or stars are nearer to the Earth, or are some stars nearer than the planets and some farther away?*

Provide children with star-shaped question cards (Resource sheet 2). Ask them to write questions they want to find answers to. Add questions to guide their learning:

Do the Moon and planets give out light?

If the Moon and planets don't give out light, how can we see them?

Only five planets can be seen with the naked eye: Mercury, Venus, Mars, Jupiter and Saturn. Why do you think this is?

Explain that the next part of the journey through space might help to answer some of their questions, but others will be answered in later lessons or through their own research.

Explain to children that their challenge is to draw a labelled diagram of the solar system, including as much information as possible. Explain that they are going to look at images of the other planets, so should make notes on their whiteboards to help them to remember important facts.

Display slide 3 from Slideshow 1 (Space Shuttle launch) and then slide 4 (labelled diagram of the solar system). Draw attention to what can be seen in the diagram.

The challenges are differentiated by the amount of information children are required to provide. Children work co-operatively but draw their own diagrams.

Challenge 1: Children draw a labelled diagram of the solar system

Ask them to use their notes to help them to draw and label a diagram of the solar system and, if they can, add some notes about the planets' movements. Print out colour copies of the Planets (Resource sheet 3, Collins Connect only).

Write this mnemonic on the board: My Very Educated Mother Just Served Up Nachos.

Ask: *How can this help you to remember the order of the planets, in order of distance from the Sun?*

Challenge 2: Children draw an annotated, labelled diagram of the solar system.

Ask the children to draw and label a diagram of the solar system showing the Sun, planets and anything else that they know is there.

Ask: *Can you show how the planets move in space? What is beyond our solar system?*

Challenge 3: Children draw an annotated, labelled diagram of the solar system, noting relative sizes of planets.

Ask the children to draw a labelled diagram of the solar system with as much information as they can about the movement of the planets and anything else in it.

Ask: *How can you show sizes of the planets in relation to one another? How can you show other things in space?*

While the children are working, group the star questions according to related ideas.

REFLECT AND REVIEW:

Ask children to leave their work on the table and look at everyone else's work. Then ask them to return to their card sort.

Ask: *Have you changed your mind about any of your ideas? What changed your mind?*

Select 'star' questions related to the lesson. Ask children if they found the answers.

They could add answers and new questions to the display as the module progresses.

EVIDENCE OF LEARNING:

Observe children as they discuss the card sort. Listen to their questions. Look at their solar system diagrams. Can children ask questions about the solar system? Can they name the planets? Do they draw spherical planets? Do their diagrams show that the planets move around the Sun? Do they evaluate the different sources of information in terms of how well they help to answer their questions?

CROSS-CURRICULAR OPPORTUNITIES:

This lesson links to Art with a study of Vincent Van Gogh's 'Starry Night', and to English with children choosing a planet to research for a class 'Explore a Planet' book (printed or electronic), including size, duration of orbit, comparison with an Earth year, duration of rotation and comparison with an Earth day.

THE EARTH AND BEYOND

LESSON 2: WHAT IS A YEAR?

LESSON SUMMARY:

In this lesson children draw a large 'plan' of the solar system and an annotated scientific diagram of the Earth's orbit which they use to explain the year, number of days in a year, leap years and how astronomers in the past used the stars as markers for the start and finish of an orbit.

By the end of this lesson children know that a year is the time a planet takes to orbit the Sun and that for the Earth this takes 365¼ days, so that every four years we have a 'leap year' with an extra day (366 days).

Preparation required: Arrange access to the playground or other large area (at least 20 metres squared) where the planets' orbits can be chalked on the ground.

Key vocabulary:

Earth, fixed stars, galaxy, Jupiter, leap year, Mars, Mercury, Milky Way, nebula, orbit, planet, Saturn, solar system, star, Sun, Neptune, Uranus, Venus, year

Resources:

A big ball of string (at least 20 metres in length), a big block of chalk or a couple of packs of chalk (different colours, if possible), eight large balls, card strips about a third A4, each with the name of a planet

National curriculum links:

Describe the movement of the Earth and other planets in the solar system relative to the Sun

Learning intention:

To use a model to describe and compare the movements of different planets in space

Scientific enquiry type:

Finding things out using a wide range of secondary sources of information

Working scientifically links:

Reporting and presenting findings from enquiries, including conclusions, causal relationships of and degree of trust in results, in oral and written forms such as displays and other presentations

Success criteria:

- I can use models and secondary sources of information to explain how the planets orbit the Sun.
- I can explain how the length of a year was decided in ancient times.
- I can explain what a leap year means and why we have them.

EXPLORE:

Ask children what they know about orbits of planets. Make a note of their ideas.

Ask: *Do all planets go round the Sun in the same direction? Are their orbits all on the same level or do they cross one another? Which planet has the shortest orbit? Which has the longest? Do they move at about the same speed? How is an orbit related to a year? Is it the same for all planets?*

Note the different ideas on the board.

ENQUIRE:

Ask children to look at their solar system diagrams from Lesson 1.

Ask: *How many planets are there? Are they still or moving? Where do they move? What do we call their path around the Sun? How long does it take the Earth to orbit the Sun? Do the other planets take the same time?*

Establish that the time the Earth takes to orbit the Sun is a year. Explain to them that they are going to make a giant solar system model and walk the orbits of planets to find out about their years.

Take the class into the playground or a large room. Mark the Sun at its centre. Choose a child to chalk the orbit of the farthest planet from the Sun (Neptune). Ask another to stand on the Sun holding the end of a piece of string while the orbiting child holds the other end, goes to the closest wall and cuts the string to that length. Ask them to tie the string to a piece of chalk and chalk a circle, keeping the string taut: the string acts as a compass.

With the class, decide how far out each planet should be, cut the string in the same way and let them chalk the other planets' orbits – if possible in different colours. Do not worry about relative distances of orbits from each other, just their order from the Sun.

Choose a different child to represent each planet; give each a ball with a planet label. Ask them to line up to begin their orbits as if on a straight track. Mark this the start point. Ask them to walk (anticlockwise) in step, as you count 'one, two, three...'. Ask the others to watch and check that the planet children stop when they get back to their markers. When the last child is back at the start, ask the class what they noticed.

Ask: *Do the planets keep together or do some overtake others? Which planets have a shorter year than the Earth? Which ones have a longer year? How did we know where each planet's orbit began and when it got back there?*

Back in the classroom, prompt children to think about an Earth year.

Ask: *How many days does it have? How do we know when a day begins and ends?*

Explain to children that astronomers in the past counted the number of sunsets from the beginning to the end of the Earth's orbit – 365, although some thought the Sun orbited the Earth.

Ask: *Astronomers couldn't mark the starting point as we did, so how did they know when the Earth had completed an orbit? What markers, that don't move, would it pass in the sky,*

Display the first slide of A fixed star (Slideshow 1). Point out the stars, in particular the named bright star. Ask children to look carefully at that same star on each of the following photos and then display the remaining slides.

Ask: *What happens? Why does the star seem to move?*

Explain to children that ancient astronomers measured and recorded the position of a star in the sky and used this as their marker. They then counted the sunsets until the star was back in exactly the same place. Inform them that an orbit takes a little longer than 365 days – 365.25 days.

Ask: *How many hours are in a quarter of a day? After four years how many extra hours would that be? What do we do about those extra 24 hours (a complete day)? What date does the extra day have every four years?* (If necessary, introduce 'leap year'.)

Explain to children their challenge is to draw diagrams of different planets' orbits and use these to explain year length. Challenges are differentiated by the detail required on the diagrams. Children work in groups but record individually.

Challenge 1: Children draw a diagram to explain the length of the Earth's orbit

Ask the children to draw a diagram of the Earth's orbit to explain what a year means and how long it is. They could add information about how a star we see in the sky seems to move over the year.

Challenge 2: Children draw a diagram of the solar system to explain the length of different planets' orbits

Ask the children to draw a diagram of the solar system to explain what a year means, how long it is and how astronomers calculated the number of days in a year. They could add information about the significance of the changing position of a star in the sky.

Challenge 3: Children draw a diagram of the solar system to explain how stars are used to calculate the length of the Earth's orbit

Ask the children to draw a diagram of the solar system to explain what a year means, how long it is, how the number of days in a year were counted and how this affects the calendar. Encourage them to add further information (for example, about how the observation of a star shows the start and end of the Earth's orbit).

REFLECT AND REVIEW:

Give children time to look at the different diagrams.

Ask: *What have you learned about the time it takes planets to orbit the Sun? How is our calendar linked to the solar system? How might you explain what you have learned today to an alien?*

Ask children to write any questions they would like answered on star cards and add them to the question board. Are there any questions that can now be answered?

EVIDENCE OF LEARNING:

Observe children as they decide where to chalk the planets' orbits. Do they relate this to what they learned in Lesson 1 (order of planets from the Sun)? Do they consider how far apart to chalk orbits? Do children know that there are 365¼ days in a year? Do they recognise that we know how many days there are in a year by counting the sunsets during the Earth's orbit? Can children explain that astronomers could tell when the Earth completed an orbit by observing and measuring the position of a star? Do they use the term 'leap year' correctly?

Key information:

The stars stay in the same places in the sky, but as the Earth orbits the Sun, their heights in the sky change each day.

THE EARTH AND BEYOND

LESSON 3: WHAT IS A DAY?

LESSON SUMMARY:

In this lesson children investigate how the Earth's rotation causes the apparent movement of the Sun across the sky. They know from previous lessons in this module that the Earth's orbit determines the length of a year and why we have leap years. By the end of this lesson they understand that the Earth's rotation causes night and day.

Preparation required: Identify a suitable place in the school grounds, facing south, for children to sketch. Back in class they add where they think the Sun will be at given times.

Key vocabulary:

axis, dawn, dusk, Earth, horizon, rotate, spin, Sun, sunrise, sunset

Resources:

Globe, poster putty, bright torch, cocktail stick, compass

Health and safety:

Ask children not to look directly at the Sun because it is so bright that it could harm our eyes, not to shine a torch directly at anyone's face and not to look at a bright torch bulb.

Key information:

The Earth takes 24 hours to rotate once on its axis (anticlockwise).

The Sun rises due east on only two days of the year: the equinoxes of March 20th/21st and September 22nd/23rd. In the UK, sunrise is almost northeast in June and almost southeast in December – a 90 degree change in direction.

National curriculum links:

Use the Earth's rotation to explain day and night and the apparent movement of the Sun across the sky

Learning intention:

To use a model or diagram to explain the effect of the Earth's rotation in space.

Scientific enquiry type:

Noticing patterns

Working scientifically links:

Reporting and presenting findings from enquiries, including conclusions, causal relationships of and degree of trust in results, in oral and written forms such as displays and other presentations

Success criteria:

- I can identify patterns in my observation of shadows.
- I can explain why the Sun appears to move across the sky from east to west.
- I can explain how night changes to day and back to night.

EXPLORE:

Inform children that they are going to look at photographs of the same place at different times of the day and notice how light and shadows change.

Display slides 1–6 of What is a day? (Slideshow 1), in sequence.

Ask: *What did you notice?*

They should notice that in the first and last pictures sunlight is weak. Introduce 'dawn' for the time around sunrise and 'dusk' for the time approaching sunset. Ask where the Sun is at these times. Establish that at dawn and dusk the Sun is low, near the horizon.

Establish that the light source in the photographs is the Sun, shadows are formed where objects block the Sun's light, and the higher the Sun is in the sky, the shorter the shadows (and vice versa). Ensure that children understand that when the Sun moves across the sky shadows change direction.

Show slide 7 from Slideshow 1. Explain that someone took photographs of the same place at different times on the same day, and then put the different positions of the Sun from all the photos onto one image to show how it moves across the sky.

Ask: *What pattern do you notice? Which Sun is at noon?*

ENQUIRE:

Leave Slide 7 on display. Put a piece of poster putty on the UK on the globe. Push the cocktail stick into it. Rotate the globe anticlockwise. Show that this is from west to east, viewed from above the North Pole.

Ask a child to shine the torch on the globe. Ask the class to notice how the stick's shadow changes as the globe turns. Ask them to imagine how the Sun would seem to move if they stood where the stick is.

Ask: *What happened to the shadow as the globe turned? Does the Sun move? What makes the shadow change?*

Ask children to repeat the globe rotation with the torch shining on it and to notice if everywhere has night at the same time.

Ask: *What happens in different countries as the Earth rotates? Is there a gradual or sudden change from night to day? Does the whole world have night or day at the same time? Why not?*

Prompt children to think about the length of a day and how it may relate to the Earth's rotation.

Ask: *How long does a complete rotation of the Earth take?*

Explain that people long ago found it useful to mark time by splitting a day into the 24 segments that we call hours, and splitting them into minutes and seconds.

Show slide 7 of Slideshow 1 again and ask children in which direction the Sun appears to move (roughly east to west).

Ask: *In which direction does the Earth rotate?*

Let a volunteer check this by turning the globe so that the Sun appears to move from east to west. Make sure that the whole class can see.

Advise children that their challenge is to draw diagrams of the position of the Sun at different times. The challenges are differentiated by difficulty of task, with Challenge 3 requiring children to use a compass and create their own map.

Challenge 1: Children identify countries that are in darkness during UK daytime

Ask the children to work as a group, using the torch and the globe to find three countries that have night while the UK has day. Ask each child to draw a diagram to show this, to annotate it and explain why it happens.

Challenge 2: Children identify the position of the Sun and shadows at different times

Ask the children to work independently. Give each child a copy of the Time of day sheet (Resource sheet 1). Ask them to follow the instructions, and to draw where they think the Sun will be at different times, and draw and label the shadow of the tree at these times.

Challenge 3: Children use a compass to work out and draw a diagram to show the position of the Sun in the sky when viewed from the same location of the school grounds at different times

Give each child a copy of the Where is the sun? sheet (Resource sheet 2). Take them to a suitable place in the school grounds facing south. Ask the children, working individually, to sketch the main objects in the scene and to include plenty of space on the ground for drawing shadows. They use a compass to help them to decide where the Sun rises. Back in class, ask them to work out the positions of the Sun in the sky at the times indicated. Encourage them to discuss how they are going to do this before drawing them on their sketches. Ask them to write a brief explanation of how they estimated the Sun's positions at different times.

REFLECT AND REVIEW:

Invite children to share their answers with the class.

Ask them to write questions they would like answered on star cards and to add them to the question board. Are there any questions there that can now be answered?

EVIDENCE OF LEARNING:

Observe children's responses throughout the lesson. Look at their diagrams. Can they describe how day and night arrive gradually rather than as a sudden dark/light/dark switch? Do they say that the Sun rises in the east and sets in the west? Can they identify why the Earth rotates anticlockwise? Do children's diagrams show how shadows change in length and direction during the day? Do they attribute change in shadow length to the Sun's height in the sky? Do they attribute change in shadow direction to the Sun's movement horizontally east to west across the sky? Can they use the model of the Sun and the Earth to support their explanations?

CROSS-CURRICULAR OPPORTUNITIES:

Link to Mathematics through measurement/time. For Art, this fits in well with Impressionist artists, particularly Monet, who painted the same place at different times of day.

THE EARTH AND BEYOND

Resources:

Large paper or polystyrene plates, permanent marker pens (different colours), fairly small but sharpened pencils, modelling clay, cereal boxes, scissors, direction compasses, watches, torches, small model figures (about 8–15 cm high

Health and safety:

Ask children never to look directly at the Sun because it is bright enough to harm their eyes, also neither to shine a torch directly at anyone's face nor look at a torch bulb when switched on.

Key information:

The pencil's shadow shortens, the nearer to noon it is. After the end of March, with the clock adjustment for British Summer Time, the shortest shadow is at 1 p.m., not 12 noon (12:00). The pencil's shadow also changes direction. The cereal box shadow clock is used only to record the length of the shadow each hour and is placed to face away from the Sun.

Children are asked to cut a 'slider' from the piece that has been cut off one end of the box. This helps them to check that the shadow is parallel with the end of the box without looking at the Sun to line it up.

© **LESSON 4: HOW DOES THE SUN HELP US TO MEASURE TIME?**

LESSON SUMMARY:

In this lesson children build on their learning about how the Earth's rotation makes the Sun appear to move across the sky. They test different types of shadow clock. Children record the position and length of a shadow and by the end of this lesson have evaluated the accuracy and potential uses of their clocks.

Preparation required: This lesson must be set up beforehand. Children need to make shadow clocks from paper plates or cereal boxes and record the positions/lengths of shadows on a previous day, as well as in this lesson (see Module Introduction).

Each group should make their shadow clock using either the Paper plate shadow clock instructions (Resource sheet 1) or the Cereal box shadow clock instructions (Resource sheet 2). Identify/arrange access to an appropriate outdoor area of the school grounds that has full sunlight all day (probably to the south of any buildings) for testing shadow clocks.

National curriculum links:

Use the idea of the Earth's rotation to explain day and night and the apparent movement of the Sun across the sky

Learning intention:

To make a shadow clock and test its accuracy

Scientific enquiry type:

Observing changes over different periods of time

Working scientifically links:

Taking measurements, using a range of scientific equipment, with increasing accuracy and precision, and taking repeat readings when appropriate

Success criteria:

- I can take accurate measurements and record them accurately.
- I can make predictions based on my measurements.
- I can test a shadow clock to check how accurate it is.

EXPLORE:

Display all the shadow clocks.

Ask: *What pattern do you notice in the shadows over the course of a day? What is happening?*

Check that children noticed that shadows get shorter towards noon and then get longer. Establish that the Sun appears to move higher in the sky towards noon and then gradually lower.

Invite a child to use the torch and a small model figure to demonstrate this. Ask the child to shine the torch onto the figure from a low level and then gradually raise the torch, keeping it trained on the figure. The shadow shortens. Reverse this process, starting with the torch high and then lowering it; the shadow lengthens.

Ask: *At what time was the shortest shadow? Was this what you expected?*

If the shortest shadow was not at 12.00, ask if they can explain why. If necessary, explain that in the UK we move clocks forward 1 hour at the end of March (the start of British Summer Time), so 'noon', when the Sun is at its highest, is at 1 p.m. We move clocks back 1 hour at the end of October.

Ask: *In what other ways do shadows change during the day? Why?*

Children should notice that the direction of the shadows changes. Remind them of their observations about how the Sun appears to move across the sky from east to west.

Invite a child to demonstrate this, using the torch and a small model figure. Ask the child to shine the torch on the figure and gradually to move it in a semi circle from one side to the other, keeping it trained on the figure.

Ask: *If we continued making measurements on the shadow clock until later in the day, what would eventually happen? Why?*

Make sure children can say that eventually the Sun would go out of sight. If necessary, reinforce that this is because the Earth rotates; use the terms 'dusk' and 'sunset'. Ask children when the Sun will be seen again after its disappearance at sunset. Use the terms 'dawn' and 'sunrise'.

Ask: *Where can you use your shadow clock? Would it work in a shady place? Does it matter which way around you place it?*

Children should recognise that position is important for the paper plate clock because otherwise the shadow could fall on any of the lines – or in between them. The position is important for the cereal box clock because the open end must face away from the Sun so that the end with the upright piece casts a shadow on the base, perpendicular to the sides.

Ask: *How can you check that your clock faces the correct direction?*

Help children to use a compass to match the direction of north/south to those they marked on the paper plate clock. For the cereal box clock line the shadow up along the 'slider'.

Take children outdoors to test each other's shadow clocks. Pair groups with other groups that made a different shadow clock. Ask each group to instruct the other on how to use the clock. Ask them to use their watches to check the shadow clocks.

Ask: *How accurately can you tell the time with the shadow clock? Is it accurate to the hour, half hour, quarter hour, for example?*

The challenges provide an opportunity for children to apply to different situations to their understanding of how shadows change during the course of a day. Challenges are differentiated by complexity of context.

Challenge 1: Children use knowledge of how the Sun casts shadows at different times of day to spot and correct mistakes in drawings

Give the children Shadow graph sheet (Resource sheet 3). Note: No shadows have been omitted so that teachers can choose which to delete.

Ask the children to correct any shadows they think have been drawn the wrong length or in the wrong direction, and to explain what was wrong and how they knew.

Challenge 2: Children use knowledge of how the Sun casts shadows at different times of day to match shadow times to their lengths and directions

Give children Shadow stick sheet (Resource sheet 4). Note: No shadows or times have been omitted so that teachers can choose which to delete. Ask them to match times of day to the shadows and to draw the missing shadows. Ask children to explain how they knew how long and in which direction to draw the shadows.

Challenge 3: Children use knowledge of shadow clocks to identify which type would make a good portable clock

Ask the children to look at the two types of shadow clocks they made and to say which would be the easier to use as a portable clock in any sunny place, and why. Ask them to draw diagrams of the clocks to help.

REFLECT AND REVIEW:

Ask: *How do shadows change during a day? What makes them change? How can this help us to tell the time? Which shadow clock would you recommend to someone who doesn't have a watch?*

Ask children to write a three-line advert to persuade someone to use their shadow clock.

Ask them to write questions they would like answered on the star cards and to add them to the question board. Are there any questions that can now be answered?

EVIDENCE OF LEARNING:

Observe and listen to children's responses. Notice how they use their observations of shadows to complete the challenges. Can they tell which shadows are made nearer to noon/later or earlier? Can they describe how the Sun's apparent movement across the sky from east to west affects directions of shadows? Can they describe how the Sun's apparent rising above the horizon and setting below it affect lengths of shadows? Can children evaluate the accuracy of shadow clocks by comparing shadow measurements with the time on a clock or watch?

If possible, arrange to construct a long-term 'human Sun clock' in the school grounds. This is a simple calendar using shadow markers, with a place for the human 'shadow stick'. The calculations are quite complex, but there are organisations that provide layout kits, for example, www. sunclocks.com.

THE EARTH AND BEYOND

LESSON 5: WHAT TIME IS IT AROUND THE WORLD?

LESSON SUMMARY:

In this lesson children use a globe and world maps to find out about world time zones and how time is linked to longitude. They build on learning from earlier lessons in this module about how the Earth's rotation makes the Sun appear to move across the sky from east to west and to become higher in the sky as noon approaches and lower towards evening. By the end of this lesson children are able to explain changes in sunrise and sunset times through the year and use a map to find times around the world.

Key vocabulary:

British Summer Time, degrees, International Date Line, longitude, meridian, rotation, sunrise, sunset, time zone

Resources:

Globes, torches, sticky tack, large map of the world that shows major cities, online maps of the world that show longitude, internet world clock

National curriculum links:

Use the idea of the Earth's rotation to explain day and night and the apparent movement of the Sun across the sky

Learning intention:

To use a model to explain why sunrise and sunset occur at different moments in time in different parts of the world

Scientific enquiry type:

Finding things out using a wide range of secondary sources of information

Working scientifically links:

Recording data and results of increasing complexity using scientific diagrams and labels, classification keys, tables, scatter graphs, and bar and line graphs

Success criteria:

- I can use lines of longitude on a map to find the time in different places around the world.
- I can put a list of cities around the world in order of when they have sunrise and begin a new day.
- I can explain how people around the world use time zones to organise their clocks and calendars.

EXPLORE:

Display What time is it where you are? (Resource sheet 1). Select children to read out the phone calls.

Ask: *Why does Sonia have breakfast as her father has lunch?*

If necessary, prompt children to think about where Sonia's father is and what difference that makes. Establish that it is not because Sonia got up late/her father is having lunch early, but because the time is different there.

Ask: *What time is it in Moscow?*

Ask children which country Moscow is in. Let them find it on a globe or on a world map .

Ask: *Why is Mike in bed? What time is it? Where is Toronto?*

Ask children to find Toronto on a map or globe.

Ask: *Where is Sydney? Is it morning, afternoon or evening there? What time is it?*

Ask children to find Sydney on a map or a globe.

When children have read the phone call to Honolulu (Hawaii), help them to find it on a map or globe.

Ask: *Is it earlier or later in Honolulu than in London?*

It might seem later because it is evening in Honolulu but morning in London.

Ask: *If it's later, why does Sonia's grandmother say, "I'll phone you tomorrow, on your birthday"?*

Children may be able to figure this out from the map or globe, especially if they have travelled a lot.

Remind them about what they learned (Lesson 3) about how the Sun appears to move across the sky because of the Earth's rotation. Ask two children to demonstrate this using a torch and globe. Ask them to watch how the patch of light moves from east to west.

Put pieces of sticky tack on London, Moscow, Toronto and Sydney on the globe.

Ask: *In which of these cities was it the latest when Sonia phoned her family?* (Sydney). That means that Sydney had sunrise first.

Shine the torch and rotate the globe so that it is sunrise in Sydney. Ask which city has sunrise next? (Moscow). Continue to rotate the globe so that Moscow then London have sunrise.

Ask children what pattern they notice. Explain, if necessary, that sunrise moves westwards.

Ask: *What about Honolulu?*

Put a piece of sticky tack on Honolulu. Continue slowly rotating the globe.

Ask: *Is it earlier or later in Honolulu or Toronto?*

Key Information:

Note that in Resource sheet 1 the time in Honolulu appears to be much later in the day than the time in the other cities, whereas in reality it is earlier because it is the day before. Children might not understand this until they explore longitude/ sunrise around the world.

Continue rotating – we're back to Sydney!

Ask: *So was it earlier or later in Honolulu than in Toronto? Was it earlier or later in Honolulu than in Sydney?*

Continue round all the cities again.

Ask: *How can we really tell?*

Establish that to know where is earlier or later we need a starting point.

ENQUIRE:

Split children into small groups. Give each group a map or globe. Explain to them that scientists numbered lines of longitude to divide the Earth into time zones. Ask them to trace some of these lines with a finger. Explain longitude.

Ask: *Can you find the starting point using the numbers on these lines?*

Point out the line marked zero and explain to children that lines of longitude are sometimes called meridians. Display the Greenwich Meridian image (Slideshow 1) and explain what this is (find out more on the Greenwich Meridian website: http://www.thegreenwichmeridian.org).

Ask children to move westwards across the map to find the highest number (180°), then to go back to zero and move eastwards to the highest number.

Ask: *What do you notice?*

They should see that 180° west is also 180° east – half a rotation of the Earth. Explain the International Date Line.

Ask if anyone can use the International Date Line to explain why 10 p.m. in Honolulu is 'earlier' than 9 a.m. in London?

Explain to children that their challenge is to find out more about the relationship between longitude and time. The challenges are differentiated by the difficulty of the task, moving from straightforward information retrieval in Challenges 1 and 2 to application of understanding in Challenge 3. Group children in threes.

Challenge 1: Children use an internet world clock and map to find out the time at places around the world

Provide the children with a world map showing major cities and countries. If cities are not marked, support this with maps of the relevant countries. Give them the World map outline (Resource sheet 2), which includes lines of longitude, and Questions sheet (Resource sheet 3). Ask the children to follow the instructions and answer the questions on their challenge sheet. Remind them to use the 24-hour clock.

Challenge 2: Children use an internet world clock and map to find out the time and longitude at places around the world

Provide the children with World map outline (Resource sheet 2) and Questions sheet (Resource sheet 4). Ask them to follow the instructions and answer the questions on their Challenge sheet. Remind them to use the 24-hour clock.

Challenge 3: Children use lines of longitude to work out the time at places around the world

Provide the children with World map outline (Resource sheet 2) and Questions sheet (Resource sheet 5). Ask them to follow the instructions and answer the questions on their Challenge sheet. Remind them to use the 24-hour clock.

REFLECT AND REVIEW:

Ask children to write a text to wish an (imaginary) friend in New York luck for a sports competition to be held at 3 p.m. on May 6th. Ask them to say what they are doing at the time the text is being written.

Ask: *By what time must you send the text to get to your friend before the competition?*

Ask children to write questions they would like answered on star cards and to add them to the question board. Are there any that can now be answered?

EVIDENCE OF LEARNING:

Observe children as they collect information about the time at different longitudes. Can they demonstrate that sunrise and sunset move gradually around the world from east to west and explain why the farther east a place is, the earlier it is? Can they explain that scientists made longitude 180° International Date Line a starting point for the date for the entire world? Can children use the world clock or longitude to check times in different cities?

Key information:

The Sun rises gradually round the world from east to west, and at noon moves westwards around the world. There comes a point where noon of one day is also noon of the next day. This point, which can be thought of as a line that stretches from the North Pole to the South Pole at 180° longitude, passes through the Pacific Ocean and was adopted as the International Date Line in 1884. It is not a completely straight line and does bend round some islands rather than cut through them, to avoid giving part of an island a different date from the rest. When ships cross the International Date Line going east they move the calendar back one day; going west they move it forward a day.

Key information:

Remind children of their learning in Maths about measuring full revolutions (rotations) and half or quarter revolutions in degrees – a full revolution being 360°.

THE EARTH AND BEYOND

LESSON 6: WHY DO WE HAVE SEASONS?

Key vocabulary:

autumn, axis, equinox, hemisphere, northern, North Pole, orbit, rotation, solstice, southern, South Pole, spring, summer, sunrise, sunset, temperature, tilt, winter

Resources:

Battery powered lanterns that shine in all directions (or torches), a globe, poster putty, materials for making a poster, piece of dowelling, small ball of modelling clay, secondary sources for research

Health and safety:

Ask children never to look directly at the Sun because it is bright enough to harm our eyes, not to shine a torch at anyone's face and not to look at a torch bulb when switched on.

Key information:

The Earth is tilted at an angle of 23.5°. This is the reason we have seasons in the UK. The tilt angles Earth's northern hemisphere towards the Sun in the summer and away from it in winter, with midway points in spring and autumn.

LESSON SUMMARY:

In this lesson children explore how the Earth's tilt on its axis causes seasonal changes and changes in daylight hours. They build on what they have learned about time in previous lessons in this module. By the end of the lesson children are able to explain that the Earth's tilt causes seasons, and how seasons in the northern hemisphere differ from those in the southern hemisphere and from those in tropical regions.

National curriculum links:

Describe the movement of the Earth, and other planets, relative to the Sun in the solar system

Learning intention:

To explain how the Earth's tilt leads to seasonal changes

Scientific enquiry type:

Observing change over time (modelled)

Working scientifically links:

Recording data and results of increasing complexity using scientific diagrams and labels, classification keys, tables, scatter graphs, and bar and line graphs

Success criteria:

- I can describe how the Earth orbits around the Sun while it is turning on its axis.
- I can explain how the tilt of the Earth's axis causes seasons.
- I can use my pictures or models to explain why a season is not exactly the same in different parts of the same hemisphere.

EXPLORE:

Display Seasonal changes sheet (Resource sheet 1) and ask children to think about the seasons we have in the UK.

Ask: *What are the seasons? What is it like in each season? What kinds of weather do you expect in spring? Summer? Autumn? Winter? How do temperatures change from one season to another? Does anything else change? Have you heard people say "the days are getting longer" or "the nights are drawing in"? What do they mean?*

Open a discussion with children and write their responses in the table.

Ask children to talk with a partner about what makes the seasons change. Invite feedback and record their ideas to refer to later. Show children the globe and ask what they notice about the position of the Earth on the stand.

Ask: *Why do globes always show the Earth tilted? Is it really tilted or is this just for models?*

Explain that scientists think another planet bumped into the Earth billions of years ago and made it tilt.

Let children look at a globe from above the North Pole and slowly rotate it is anticlockwise. Point out that the Earth and other planets rotate around their poles. Push a piece of dowelling through a small ball of modelling clay and explain that the dowelling shows the line of the Earth's axis (an imaginary line running through the centre of the Earth between the two poles), but point out that there nothing really runs through the Earth.

Point to the top of the globe and ask what this part of the Earth is called. Point to the South Pole and the equator. Show them the northern and southern hemispheres. Ask a child to mark the UK with a piece of sticky tack.

Ask: *Which hemisphere is the UK in?*

ENQUIRE:

Explain to children that we can use a globe to represent the Earth and a lantern or torch to represent the Sun. Explain that the model is not to scale – if a globe represents the Earth, then whatever we use to represent the Sun should be bigger than the room.

Shine a lantern or torch on the globe to represent sunlight.

Ask: *Do all parts of the globe get the same amount of light? What season would we have when the northern hemisphere is tilted towards the Sun?*

Move the globe to orbit the lantern or torch 'Sun'. Stop when the southern hemisphere tilts towards the Sun. If using a torch, turn it to shine on the globe.

Ask: *Which hemisphere is now tilted towards the Sun? What season will it be now in the northern hemisphere?*

Show the class Seasons animation (Animation 1), which provides a visual explanation of our seasons.

Remind children of some of their observations about the seasons that you recorded on the chart (Resource sheet 1) at the start of the lesson. Explain to them that their challenge is to use what they know about the Earth's tilt to describe and explain differences between the seasons. The challenges are differentiated by the number of seasons children are required to describe and explain.

Challenge 1: Children make a poster to show the position of the Earth and the Sun during winter in the UK

Ask the children to model the Earth's rotation (remembering the tilt) and to draw a poster-sized, labelled scientific drawing with the Sun and the Earth to show winter in the UK. Ask them to name three other countries that would have winter at the same time and to say how they know.

Encourage the children to attempt to describe the seasons that Australia, Ghana and Antarctica would have at the same time. They can use a different copy of the blank Seasons sheet (Resource sheet 1) for each country.

Challenge 2: Children make a poster to show the position of the Earth and the Sun during summer and winter in the UK

Ask the children to model the Earth's rotation (remembering the tilt) and to draw a poster-sized, labelled scientific drawing with the Sun and the Earth in two different positions in the Earth's orbit; one for summer and one for winter in the UK. Ask the children to use the terms 'summer' and 'winter', 'Earth's orbit', 'Earth's tilt' and 'the Sun'.

Encourage the children to attempt to describe the seasons that Australia, Ghana and Antarctica would have, firstly when the UK has summer and then when the UK has winter. Ask them to say whether the season is the same in each of the three countries – and why. Give the children copies of the Seasons table (Resource sheet 1) so that they can record the seasons in the countries. Different children can focus on different countries.

Challenge 3: Children make a poster to show the position of the Earth and the Sun during all four seasons in the UK

Ask the children to model the Earth's rotation (remembering the tilt) and to draw a poster-sized, labelled scientific drawing that includes the Sun and the Earth in the four different positions that the Earth will be in its orbit when it is spring, summer, autumn and winter in the UK. Explain to the children that they should use the names of the seasons and the terms 'Earth's orbit', 'Earth's tilt', 'the Sun', 'solstice' and 'equinox'. Ask them to choose a season in the UK and to investigate (using secondary sources to check their ideas) whether this would be the same for the entire northern hemisphere. Ask them to record their findings.

If there is time, the children could also describe the season a country in the southern hemisphere would have when the UK has winter, and find out if this is the same for other countries in the southern hemisphere (using secondary sources to check their ideas). These tasks could be shared among the group.

REFLECT AND REVIEW:

Ask children to present their findings orally using their models and drawings to help. Return to their ideas about the seasons from the start of the lesson (including the completed table on Resource sheet 1) and ask if their investigations have helped to explain any seasonal changes – and if so, how?

Ask them to write questions they would like answered on the star cards and to add them to the question board. Are there any questions that can now be answered?

EVIDENCE OF LEARNING:

Can children describe the different seasons in the UK in terms of day length and temperature changes? Do they show in their drawings, modelling and presentations that the Earth's axis tilts? Can they use a model to explain how the tilt of the axis causes the seasons?

As a possible extension to this lesson, challenge children to find out why some countries don't have the same four seasons as the UK: for example, Pakistan, Ecuador, Indonesia and Republic of Congo.

THE EARTH AND BEYOND

LESSON 7: WHAT ARE OUR CONCLUSIONS ABOUT SUNRISE AND SUNSET TIMES?

Key vocabulary:

Arctic, Antarctic, autumn, axis, equator, equinox, hemisphere, northern, North Pole, orbit, rotation, solstice, southern, South Pole, spring, summer, sunrise, sunset, temperature, tilt, winter

Resources:

Globe, torch, maps and atlases of the UK and the world, access to the internet for further research

Health and safety:

Remind children of safe practices when using the internet.

Key information:

The Earth is tilted at an angle of 23.5°. This tilt angles either the Earth's northern or southern hemisphere towards the Sun. The hemisphere tilted towards the Sun has summer and the other, being tilted away from the Sun, has winter. As a result there are changes in the hours of daylight. Day and night are the same lengths at the equinoxes: September 22nd and March 20th. The solstices on June 21st and December 21st are the 'longest' and 'shortest' days (those with the longest or shortest period of daylight). Arctic and Antarctic regions have periods of 24-hour, or almost 24-hour, daylight or night. Days and nights in tropical and equatorial regions are equal, or almost equal, in length throughout the year.

LESSON SUMMARY:

This lesson develops children's learning on time and seasons from the previous lessons in this module, through investigating and explaining changes in the times of sunrise and sunset in different parts of the UK and different parts of the world. By the end of the lesson children are able to explain how the Earth's tilt affects the times of sunrise and sunset in different places at different times of the year.

National curriculum links:

Use the idea of the Earth's rotation to explain day and night and the apparent movement of the Sun across the sky

Learning intention:

To be able to explain how the Earth's tilt affects the times of sunrise and sunset in different places at different times of the year

Scientific enquiry type:

Finding things out using a wide range of secondary sources of information

Working scientifically links:

Identifying scientific evidence that has been used to support or refute ideas or arguments

Success criteria:

- I can use secondary sources to find out the times of sunrise and sunset in different places.
- I can record my results in a line graph that shows a gradual change over time.
- I can explain how the tilt of the Earth's axis causes changes in the hours of daylight in different seasons.

EXPLORE:

Remind children of some 'seasonal' sayings that were mentioned in Lesson 5: "the days are getting longer" and "the nights are drawing in", and what people mean by these. Children should be able to explain that the days are still 24 hours long but that the hours of sunlight and darkness change.

Ask: *When are nights at their shortest? When are nights at their longest? When are they about the same length as daytime? What makes the night begin earlier? What causes the time of sunset to change?*

Children learned in Lesson 6 that the Earth's tilt on its axis causes the changes in seasons because either the northern or southern hemisphere is angled towards or away from the Sun. Remind them of the meanings of northern and southern hemisphere. Ask two children to use the globe and a torch to remind the class how sunrise spreads round the world from the east. Ask another pair of children to demonstrate and remind the class how the Earth's tilt affects this.

ENQUIRE:

Give each child the Sunrise and sunset table (Resource sheet 1) and Sunrise and sunset graph template (Resource sheet 2). Split the class into groups of three and ask each group to use secondary sources of information to find out how the times of sunrise and sunset change during a month in their town. Different groups could collect results for a different month: March, June, September and December (the months of the solstices, and equinoxes). Explain why a line graph would be good for showing the times of both sunrise and sunset each day, and help them to prepare their graphs.

Invite the class to share their results. Ask what patterns they notice in each group's results. They should notice turning points at the solstices and equinoxes. Ask what conclusions they can draw. Explain to children that their challenge is to discuss their findings with a partner, to predict daylight hours for other places in the world and to find data to support or refute their predictions. The challenges are differentiated by the amount of data children are required to analyse. They work in pairs but each writes an individual record of the answers that they find.

Challenge 1: Children investigate daylight hours in Cape Town

Provide the children with the Sunrise and sunset table (Resource sheet 1), the Sunrise and sunset graph template (Resource sheet 2) and the Sunrise and sunset investigation sheet (Resource sheet 3). Ask the children if they think Cape Town in South Africa will have sunrise and sunset at similar times to their own town in December. Help them to use the globe and torch to explore this before making and explaining their prediction.

Now ask the children to plan an investigation into sunrise and sunset times in Cape Town using the investigation sheet (Resource sheet 3), and to collect and record their results. Advise them that they can use the Sunrise and sunset table and graph templates for recording and presenting the results. Ask the children to compare the results for Cape Town with the graph of their own town in December and to draw a conclusion. Encourage them to raise questions about sunrise and sunsets in Cape Town at other times of the year.

Challenge 2: Children investigate daylight hours in Orkney and the Scilly Isles

Provide the children with Resource sheets 1–3. Ask them to find Orkney and then the Scilly Isles on a map of the UK, then ask them to discuss and predict whether these places will have the same times of sunset and sunrise in the summer, and why they have made these predictions. Explain to the children that they can use any of the equipment in the classroom to help them with their predictions.

Ask the children to plan an investigation into sunrise and sunset times in Orkney and the Scilly Isles, using Resource sheet 3 to help collect and record their results and draw a conclusion. The group could either share the whole task, or perhaps half the group could record the findings for Orkney and the other half records the findings for the Scilly Isles.

Challenge 3: Children investigate daylight hours in Reykjavik and Quito

Provide children with Resource sheets 1–3. Ask them to predict whether Reykjavík in Iceland or Quito in Ecuador will have the most hours of daylight in their main summer month, and discuss how they can find out using secondary sources. Advise children that they can use any of the equipment in the classroom to help them with their prediction.

Observe the children's planning and, if they assume that the main summer month for both places is the same, ask them to check this. They should plan how to record the data they collect and the best way to present it. The group could either share the task, or perhaps suggest that half the group records the findings for Quito and the other half records the findings for Ecuador.

REFLECT AND REVIEW:

Invite someone from each group to read out their question and to give a brief summary of their findings and their conclusion. Ask if their results made them ask questions and, if so, how they might find the answers.

Ask: *Were any of your results surprising? If so, how did you explain these surprises?*

Children may, for example, be surprised to find that Reykjavík has more hours of sunshine in summer than Quito. The results could be presented in a class book (or e-book) on sunrise and sunset.

Ask children to write any questions they would like answered on the star cards and to add them to the question board. Are there any questions there already that can now be answered?

EVIDENCE OF LEARNING:

Look at children's investigation reports. Can children use the internet to find sunrise and sunset times for a place in the UK or somewhere else in the world? Can they use a line graph to record and compare sunrise and sunset times over the course of a month? Can children explain why some places have more hours of sunlight in the summer than others?

CROSS-CURRICULAR OPPORTUNITIES:

This lesson can be linked to Geography, with children recognising and naming continents and countries, and also understanding climate and weather.

The measurement of time and presentation of data in charts and graphs can be linked to Mathematics.

THE EARTH AND BEYOND

LESSON 8: WHY DOES THE MOON CHANGE SHAPE?

LESSON SUMMARY:

In this lesson the children use their Moon diaries as a source of information to investigate how the Moon appears to change shape over a month. By the end of the lesson the children are able to explain that the Moon looks as if it changes shape because, although half of it is always illuminated by the Sun, we can't always see the entire illuminated half from the Earth.

Preparation required: Children need their Moon diaries, which they will have completed prior to the start of this lesson using Moon diary (Resource sheet 1) to sketch the Moon's shapes. For this lesson you also need to prepare a large ball by covering half of it with black plastic and half with a white plastic bag. Cut out (or use ready-made) pale (for example, yellow) circles to create a 'Moon phase' frieze along a border strip of dark paper along the top of a display board – if possible, the frieze could go right around the classroom.

Key vocabulary:

crescent, gibbous, orbit, the Earth, Full Moon, illuminate, lunar month, Moon, New Moon, reflect, waning, waxing

Resources:

A border strip of dark paper for a 'Moon phase' frieze, at least 30 circles for cutting out Moon shapes, chalks, a big ball half covered with black plastic and half covered with a white plastic bag, a piece of black sugar paper per child with a circle drawn in the middle of each sheet in white chalk (number these sheets individually from 1 up to the number of children in the class), access to the internet to check online calendars, a calendar for the month ahead

Key information:

The phases of the Moon that we see when we look at the night sky are shaped the way they are because from Earth we may only be able to see a portion of the Moon that is illuminated by the Sun. In space the Sun illuminates half of the Moon at any one time, but from Earth we can't always see the entire illuminated half. This is because both the Moon's orbit of the Earth and its rotation on its axis take the same length of time.

National curriculum links:

Describe the movement of the Moon relative to the Earth

Learning intention:

To identify the phases of the Moon and explain why these occur

Scientific enquiry type:

Observing changes over different periods of time

Working scientifically links:

Using test results to make predictions to set up further comparative and fair tests

Success criteria:

• I can name the phases of the Moon.
• I can explain why the Moon appears to change shape.
• I can say how long the Moon takes to orbit the Earth and how the calendar is linked to this.

EXPLORE:

Ask children to draw on mini whiteboards all the different shapes of the Moon that they have observed in their diaries. Ask them to talk to a partner about why they think the Moon appears to change shape and to write down their ideas.

Display The phases of the Moon (Resource sheet 2) and ask children to look at their own drawings and to identify any of the phases that they have drawn. They should notice which way around a crescent or gibbous Moon faces and how they appear in sequence. Ask them to count the days the full sequence takes (it is about 29 days, or more accurately, 29.5 days).

Ask: *How does this fit in with the calendar?*

Explain that a lunar month is the length of time it takes to progress through a complete sequence of the Moon phases, from one New Moon. Explain to them that the word 'month' comes from 'Moon'.

Ask: *How many months are there in a year? Do they all have 29.5 days?*

Explain that the calendar we use is divided into 12 months that have different numbers of days so that it fits the Earth's orbit of 1 year, but that some religions are based on lunar calendars.

ENQUIRE:

Seat children in a circle with the prepared black/white ball in the centre. Ask them to use chalk to draw on their piece of black paper just the white portion of the ball that they can see from where they are sitting. Point out that some children will see no white so they should draw nothing!

Then ask children to stand up in their circle, holding up their drawings. They should see the phases of the Moon as they look at the drawings. Ask children what phase of the Moon their picture shows.

Ask: *What fraction of the Moon is illuminated by the Sun? Why do we sometimes not see the Moon even if there are no clouds in the sky?*

Show Why does the Moon look like it changes? (Animation 1), which explains to the children what they have just drawn. Stop the animation after the first complete sequence of the phases of the Moon and ask children what shapes have they seen. They should identify Full Moon and New Moon.

Explain to children that as the Moon starts to be visible it is called waxing, and when it has passed through being a Full Moon and smaller portions of it are visible, the Moon is said to be waning.

Prompt children to think back to their explanations about why the Moon changes shape (from the Explore part of the lesson) and ask whether they have changed their minds and, if so, why.

Next, display the Ted the angler slideshow (Slideshow 1). Explain to them that Ted thinks that the Moon changes shape because the Earth blocks the light from the Sun.

Ask: *Can we find out if this is true?*

Explain that Ted needs to know whether there is a sequence to the patterns of the Moon's shapes, as he likes to go fishing at night, but prefers to go if it is a Full, or nearly Full, Moon so that he can see more easily. Explain to children that they are going to look at the phases of the Moon to find out when the Moon will be Full or nearly Full, to help Ted to plan his night fishing trips when there is moonlight.

The challenges are all related to this scenario and are differentiated by the degree of difficulty posed by investigating different numbers of Moon phases and the depth of explanation required. Children work in pairs.

Challenge 1: Children use cut out shapes to sequence phases of the Moon

Give each group the Moon phases sheet (Resource sheet 3), which contains pictures of the Moon in different phases and separately lists the different phase names. Ask them to cut out the pictures and the names, and to match them and put them in the right order.

Challenge 2: Children sequence and annotate the phases of the Moon

Give each group the Moon phases sheet (Resource sheet 3) and a calendar for the month ahead. Ask the children to cut out the main phases of the Moon and put them in the correct order, using their Moon diaries to help. Explain to them that they should then annotate the calendar for the coming month and check the Moon phases on the internet (for example, on www.calendar365.com).

Challenge 3: Children make a labelled diagram of the phases of the Moon

Give the children the Moon phases diagrams (Resource sheet 2) and ask them to make labelled drawings of the eight phases of the Moon in the correct order. Explain that they should then work out when in the next month the various phases of the Moon will occur. They can use the internet (for example, www.calendar365.com) to help them. Advise children that they should then write a simple explanation of the phases of the Moon to help Ted to plan his fishing trips.

REFLECT AND REVIEW:

Ask the class what they will feed back to Ted on their findings.

Ask: *What is happening to the Moon? Would Ted be able to plan his night fishing in advance? When would be the next time he could go?*

Show all children the www.calendar365.com website and check when Ted's next the fishing trip could be. Check the star question board to ensure that any questions posted there during the course of the module can now be answered. Remind children to continue with their Moon diaries so that a series of months can be displayed on a strip of dark paper along the top of a display board – and perhaps all the way around the classroom!

EVIDENCE OF LEARNING:

Look at children's diagrams and listen to the answers they provide for Ted. Do they sequence the phases of the Moon correctly? Do children explain that we see the Moon because it is illuminated by the Sun? Can they explain that the Moon seems to change shape only because we can't see all of the part that the Sun illuminates? Can children use what they have learned to help them to predict the Full Moon for the next month?

CROSS-CURRICULAR OPPORTUNITIES:

This lesson can link to Religious Education with a focus on the calendar of any religion being studied and an exploration into how and why it doesn't keep pace with the Universal Calendar. Children could explore the changing dates of religious festivals, for example: Purim (Judaism), Ramadan (Islam), Easter (Christianity), Divali (Sikhism and Hinduism) and also Buddhist and Humanist festivals.

In History, children could explore the calendar used by the ancient Romans and how it has influenced the modern calendar.

GLOSSARY

GLOSSARY

abdomen   back part of an insect's body

adult   an animal that has reached maturity

ageing   the process of getting older

amphibian   cold-blooded vertebrate animal, that lives in water or on land but must return to the water to reproduce

animal   a living organism

antennae   two long, thin body parts which are used to feel, and are attached to the head of insects and some sea animals

asexual   an organism which can reproduce by itself

bat   a mainly nocturnal mouse-like winged mammal

berry   a small edible fleshy fruit

bird   warm-blooded animal that has feathers and lays eggs with hard shells

birth   the process of bearing young; when a baby animal comes out of an egg or out of its mother

blue tit   a small songbird

blue whale   a mottled bluish-grey whale

breeding   the mating of animals and the production of offspring

brood   a family of birds or other young animals produced at one hatching or birth

bulb   rounded part of a plant usually formed underground, which grows into a new plant in the growing season

bumblebee   a large bee with a loud hum

butterfly   a nectar-feeding insect with two pairs of large, usually brightly coloured wings that are covered with microscopic scales

calf   the offspring of certain animals, for example cows

caribou   a deer of the tundra and subarctic regions of Eurasia and North America, both sexes of which have large branching antlers

carpel   female part of a flowering plant, which contains an ovary, style and stigma

clutch   a group of eggs fertilised at the same time and laid in a single session; brood of chicks

cocoon   a silky case spun by the larvae of many insects for protection

cold-blooded   having a body temperature that changes with the surroundings

compare   estimate, measure, or note the similarity or dissimilarity between itemscontrast look at why items differ

criteria   a principle or standard by which something may be judged or decided

crop   produce of cultivated plants such as vegetables and fruit; the yield of such produce

cropping   to collect the produce from the land or plant on which it has grown

cubs   the young of certain animals, for example bears and lions

cutting   a method of propagation in which a part of a plant is used to form its own roots

death   the final stage in the life cycle; the end of the life of a person or organism

dragonfly   a long-bodied predatory insect

egg   an oval or round object laid by a female bird, reptile, fish, or invertebrate, usually containing a developing embryo

elephant   a heavy, plant eating mammal with a long trunk

emperor penguin   the largest penguin, with a yellow patch on each side of the head

endangered   a species seriously at risk of extinction

environment   the surroundings in which a person, animal, or plant lives

evolution   the process by which different kinds of living organism are believed to have developed from earlier forms during the history of the earth

extinction   the process of becoming extinct

female   denotes the sex that can bear offspring or produce eggs following fertilisation

fertilise   cause an egg, female animal or plant to create a new organism by introducing male reproductive material

fertilisation   the joining of a male reproductive cell with a female reproductive cell to produce a new organism

fledge   to bring up a young bird; develop wing feathers that are large enough for flight

flower   the reproductive structure in flowering plants

foal   the offspring of an animal, for example a horse

frog   a small, squat amphibian with legs specialised for hopping

fruit   part of a plant where seeds develop

gas   state of matter where particles move about freely with no fixed shape or volume

gender   the range of physical, biological, mental and behavioral characteristics pertaining to and differentiating between masculinity and femininity

genetic   relating to genes or heredity

giant panda   a large bear-like mammal with characteristic black and white markings

gills   respiratory organ in aquatic animals

glut   an excessive amount of something, for example in the production of a crop

growth   the process of growing; an increase in size or number

habitat   a place where an animal finds the things it needs to live – food, water, and shelter

hatch   cause a young animal to emerge from an egg

head   the upper part of the human body, or the front or upper part of the body of an animal, typically separated from the rest of the body by a neck

hedgehog   a small nocturnal mammal with a spiny coat and short legs

hibernate   to spend the winter in a deep sleep

hip   a berry-like, brightly coloured fruit of a rose plant

humpback whale   a baleen whale which has a hump (instead of a dorsal fin) and long white flippers

insect   small living thing that usually has a three-part body, three pairs of legs and two pairs of wings

insoluble   unable to be dissolved

ladybird   a small beetle with a domed back which is typically red or yellow with black spots

larva   immature form of an animal, especially one that differs greatly from the adult and forms the stage between egg and pupa, for example a caterpillar

leaf (plural **leaves**)   the usually flat part of a plant that makes food for the plant

life cycle   the series of changes occurring in an animal or plant

liquid   having a consistency like that of water or oil, i.e. flowing freely but of constant volume

male   denotes the sex that can fertilise female gametes

mammal   warm-blooded animal that is covered in hair or fur. The female gives birth to live young and feeds her babies on milk from her own body

marsupial   a mammal who is born incompletely developed and is carried and suckled in a pouch on the mother's belly

material   the matter from which a thing is or can be made

metamorphosis   the process of change, for example from a caterpillar to a butterfly

migration   the movement of an animal from one habitat or region to another

monarch butterfly   large migratory orange and black butterfly that occurs mainly in North America

mountain gorilla   a powerfully built great ape with a large head and short neck. It is the largest living primate

navigate   plan and direct the course of a form of transport, especially by using instruments or maps

newt   small slender amphibian

nocturnal   active at night

opaque   not able to be seen through; not transparent

organs   tissues that work together to form a specific function

osprey   a large fish-eating bird of prey with long, narrow wings

ostrich   largest living bird, flightless and swift-running with a long neck and long legs

peregrine   falcon bird of prey, a powerful falcon found on most continents

polar bear   a large white Arctic bear

pollen   fine yellow powder made by the anthers of flowering plants to help them make new plants

pollinator   animal which transfers pollen from the anther to the stigma of a flower

pollination   the movement of pollen from the anther to the stigma of a plant so that new seeds will be produced

predator   an animal that naturally preys on others

prey   an animal that is hunted and killed by another for food

produce   agricultural products, for example farm produce

propagation   the act of producing new plants without seeds

property (plural **properties**)   a characteristic of a material

pupa   an insect in its inactive immature form between larva and adult, for example a chrysalis

pups   the young of certain animals, for example dogs or seals

reproduction   the process by which a new organism is produced

root   part of a plant that grows into the ground, anchors the plant and takes in water and nutrition

rhizome   a continuously growing underground stem

runner   a horizontal stem that grows along the soil's surface and propagates by producing roots and shoots

salamander   a newt-like amphibian

seed   small part that is made by flowers and from which a new plant can grow

seed-head   part of a flower or fruit that contains seeds

stamen   male part of a flower made up of a filament and an anther, which makes pollen

seed dispersal   the movement of seeds away from the parent plant

solid   something that is hard or firm, holds its shape and can be measured

soluble   able to be dissolved, especially in water

stem   the part of a plant that supports the branches, leaves and flowers. It helps to take water and nutrients around the plant

swift   fast-flying bird with long, slender wings and a resemblance to a swallow

tadpole   larva of an amphibian such as a frog, toad, newt, or salamander

thorax   middle part of an insect's body

thrush   small or medium-sized loud songbird, typically with a brown back and spotted breast

toad   a tail-less amphibian with a short body and legs

transparent   see-through

tree frog   typically small and brightly coloured frog found in trees

tuber   a fleshy underground stem or root

wildebeest   a large dark antelope with a long head, a beard and mane, and a sloping back (also called a gnu)

yield   the amount of produce harvested from a crop

Collins are proud to support the work of The Association for Science Education

The *Association* for **Science Education**
Promoting Excellence in Science Teaching and Learning

The ASE is the largest subject association in the UK for teachers of science.
We're a powerful force to promote excellence in science teaching and learning.

Join ASE today ...

ASE membership helps you bring science to life with innovative resources and expert advice that save you time, build your confidence and inspire your pupils.

Why you should join ...

We help you get the basics right! Not every primary teacher has a science background, and teaching science can be a complex process. ASE journals and resources are written by some of the most exciting and experienced science educators in the UK. Our materials help you to master the principles you need to teach, and to understand how young pupils build their knowledge and understanding of science.

We help you excel as a teacher.

ASE membership helps you build your skills as a teacher of science so that you can teach with confidence, generate exciting activities and inspire your pupils to explore and investigate.

Science doesn't have to be difficult – for you or your pupils. We help you to understand science – and how best to teach it.

For more details and how to join, visit www.ase.org.uk/membership/membership-category/primary/

Photo Acknowledgements

p178, left: serjio74/Shutterstock; p178, right: Brian Kinney; p189 and p305: Collins Bartholomew Ltd 2014; p210, top left: stockyimages/Shutterstock; p210, top middle: wacpan/ Alamy; p210, top right: Testi/Shutterstock; p210, middle left: lightwavemedia/Shutterstock; p210, middle: Tetra Images/Alamy; p210, middle right: FogStock/Alamy; p210, bottom left: Tetra Images/Alamy; p210, bottom middle: Hongqi Zhang/Alamy; p210, bottom right: Veronica Louro/Shutterstock; p212, top left: kostudio/Shutterstock; p212, top middle: Minerva Studio/ Shutterstock; p212, top right: Paul Orr/Shutterstock; p212, middle left: Daniel Kaesler/Alamy; p212, middle: Smirnova Irina/Shutterstock; p212, middle right: MIXA/Alamy; p212, bottom left: Andrey Bandurenko/Alamy; p212, bottom middle: MBI/Alamy; p212, bottom right: MBI/Alamy; p233, top: Yeko Photo Studio/ Shutterstock; p233, middle: mimagephotography/Shutterstock; p233, bottom: Wong Yu Liang/Shutterstock